HISTORICAL AND PHILOSOPHICAL PERSPECTIVES ON BIOMEDICAL ETHICS

New biomedical technologies and decisions require critical debate on matters as fundamental as how and what we eat, how we maintain health and how we die. This book addresses the question of an historic change from medical paternalism to patient autonomy in matters of health, or 'from medical ethics to bioethics'.

Written by authors with academic backgrounds in medicine, history, and philosophy, the contributions to this volume unlock the study of twentieth-century biomedical ethics in its social, political, legal, economic and clinical dimensions. They explore the key issues of professional self-regulation, costs of health, informed consent in medical practice and clinical research, autonomy in end of life decisions, and genetic engineering. Many of the chapters deal with German themes, giving readers a rare chance to compare the familiar with historical developments in a sister European nation.

ASHGATE STUDIES IN APPLIED ETHICS

Scandals in medical research and practice; physicians unsure how to manage new powers to postpone death and reshape life; business people operating in a world with few borders; damage to the environment; concern with animal welfare – all have prompted an international demand for ethical standards which go beyond matters of personal taste and opinion.

The *Ashgate Studies in Applied Ethics* series presents leading international research on the most topical areas of applied and professional ethics. Focusing on professional, business, environmental, medical and bio-ethics, the series draws from many diverse, interdisciplinary perspectives including: philosophical, historical, legal, medical, environmental and sociological. Exploring the intersection of theory and practice, books in this series will prove of particular value to researchers, students, and practitioners worldwide.

Series Editors:

Historical and Philosophical Perspectives on Biomedical Ethics

From Paternalism to Autonomy?

Edited by

ANDREAS-HOLGER MAEHLE
University of Durham

JOHANNA GEYER-KORDESCH
University of Glasgow

LONDON AND NEW YORK

First published 2002 by Ashgate Publishing

Reissued 2018 by Routledge
4 Park Square, Milton Park, Abingdon, Oxon OX14 4RN
605 Third Avenue, New York, NY 10017

Routledge is an imprint of the Taylor & Francis Group, an informa business

Publisher's Note
The publisher has gone to great lengths to ensure the quality of this reprint but points out that some imperfections in the original copies may be apparent.

A Library of Congress record exists under LC control number: 2001053627

ISBN 13: 978-1-138-73504-0 (hbk)
ISBN 13: 978-1-138-73498-2 (pbk)
ISBN 13: 978-1-315-18677-1 (ebk)

Contents

List of Figures and Tables

Notes on Contributors and Editors

David E. Cooper is Professor of Philosophy at the University of Durham, UK. His most recent books are *World Philosophies* (1996), *Spirit of the Environment* (editor, 1999), *Existentialism: A Reconstruction* (revised edition, 2000), and *The Measure of Things* (forthcoming, 2002).

Johanna Geyer-Kordesch is Research Professor for European Natural History and Medicine at the University of Glasgow, UK. From 1990 to 2001 she was Director of the Wellcome Unit for the History of Medicine, Glasgow University. Her most recent books include a two-volume history of the Royal College of Physicians and Surgeons of Glasgow, *Physicians and Surgeons in Glasgow* (with Fiona Macdonald, 1999) and *The Shaping of the Medical Profession* (with Andrew Hull, 1999), and a study of the life and work of Georg Ernst Stahl, *Pietismus, Medizin und Aufklärung in Preußen im 18. Jahrhundert* (2000). She has edited, with Andrew Wear and Roger French, *Doctors and Ethics: The Earlier Historical Setting of Professional Ethics* (1993).

Bryan Jennett, CBE, MD, FRCS, is Emeritus Professor of Neurosurgery at the Institute of Neurological Sciences, University of Glasgow, UK. His books include *An Introduction to Neurosurgery* (5th edition, 1994), *High Technology Medicine: Benefits and Burdens* (new edition, 1986), and *The Vegetative State: Medical Facts, Ethical and Legal Dilemmas* (2002).

Susan L. Lowe teaches in the Department of Philosophy at the University of Durham, UK.

Andreas-Holger Maehle, Dr.med., PhD, is Professor of History of Medicine and Medical Ethics at the University of Durham, UK. His recent books are *Kritik und Verteidigung des Tierversuchs: Die Anfänge der Diskussion im 17. und 18. Jahrhundert* (1992) and *Drugs on Trial: Experimental Pharmacology and Therapeutic Innovation in the Eighteenth Century* (1999).

Andrew A.G. Morrice is a General Practitioner. He was awarded an MD by the University of London for his thesis '"Honour and Interests": Medical Ethics in Britain, and the Work of the British Medical Association's Central Ethical Committee, 1902–1939' (1999).

Cay-Rüdiger Prüll, PD, Dr.med., MA, is Wellcome Senior Research Associate in the History of Medicine at the University of Durham, UK. He has edited *Pathology in the 19th and 20th Centuries* (in collaboration with John Woodward, 1998) and *Traditions of Pathology in Western Europe* (forthcoming, 2002) and is the author of *Medizin am Toten oder Lebenden? Pathologie in Berlin und London 1900–1945* (forthcoming, 2002).

Lutz D.H. Sauerteig, Dr.phil., is a Medical Historian at the Institute for the History of Medicine, University of Freiburg, Germany. He is the author of *Krankheit, Sexualität, Gesellschaft: Geschlechtskrankheiten und Gesundheitspolitik in Deutschland im 19. und frühen 20. Jahrhundert* (1999).

Marianne Sinn studied medicine at the University of Freiburg, Germany, and has recently completed her MD thesis on informed consent in German surgery 1900–33.

Ulrich Tröhler, Dr.med., PhD, is Professor of History of Medicine and Chairman of the Institute for the History of Medicine, University of Freiburg, Germany, and of the Institute for the History and Epistemology of Medicine, University of Basle, Switzerland. His most recent books are *Ethics Codes in Medicine: Foundations and Achievements since 1947* (editor, with Stella Reiter-Theil, 1998) and *'To Improve the Evidence of Medicine': The 18th Century British Origins of a Critical Approach* (2000).

Acknowledgements

Several of the essays in this volume were first presented as papers at an interdisciplinary conference on the theme 'From Medical Ethics to Bioethics', held at the University of Durham, UK, in September 1998. We would like to thank the Wellcome Trust for its generous sponsorship of this event. Many thanks also to Mrs Rae McBain, Wellcome Unit for the History of Medicine, University of Glasgow, for secretarial help in preparing the texts.

Andreas-Holger Maehle, Durham
Johanna Geyer-Kordesch, Glasgow

Introduction

Andreas-Holger Maehle and Johanna Geyer-Kordesch

New technologies create new ethical dilemmas. This is true not only of today, but of the past. What may not keep pace with the rapid quest for medical advance is the debate about it. This debate we cannot avoid, nor can it be left to the experts. Health and how we attain it implies choice and it also implies that we know what the implications of whatever choices made will be, for us and for others.

This idea is at the core of modern medical ethics. No matter how complex the science, we still need to know what its practical outcomes are. But in this lies a catch, both that of adequate explanation and of honesty. Too many statements in newspapers show scientists placating the public. They invoke a future in which disease or hunger or deformity is eradicated. But on what terms? And who profits? The honesty that we must seek is that of an open discussion. Neither what happens to our bodies nor to our natural environment can be left to the experts, not because we don't need their knowledge, but because we need their honesty and a mutual responsibility for communal welfare. This is distinctly separate from the Faustian bargain of what medicine can do in terms of science. Much is possible but not all should find our approval. No one can really claim, for example, that what is good for the consumer societies of the West is to be the template for developments elsewhere. And yet the pace of our scientific know-how, coupled with monetary investment, make medical discoveries stock market favourites. Academics turned businessmen may be tempted to leave the disinterested values of research and scholarship behind. Profit and private money fuel an autonomy that leaves an old association between public duty and the public good, in which Victorian paternalism was steeped, behind. Have we progressed?

Reflections on the past, particularly the analysis of how experimentation in medicine (on humans and animals) shaped debate, is fundamental to our examining our position today. The chapters of this book show different patterns emerging and a slow uptake in public discussion. Many of the contentious issues have been reserved to internal debates within the profession. Medicine has always been jealous of its expertise, perhaps rightly so, in view of the oft cited view that lay men or women cannot understand the science. They may well not. But as the history of medicine demonstrates, few patients were informed that they were material for trials. This was before Nuremberg, and it took repugnant atrocities, in which medicine was implicated, to set in motion the large number of codes that sought to define the balance between potential harm and the inevitable risk research on humans incurs. Yet these codes are not definitive. They seem, in their very multitude, to produce vagueness, a condition ideal for exploiting the unwary with platitudes.

Morality seems moribund compared to the glories of futurism in science. The scepticism voiced by Mary Shelley in her tale of unlimited scientific dreams and their consequences come to us as a parable, a signal for the rightness of asking about limitation, of the intuitive validity of unease. A cautionary tale like *Frankenstein* allows the sceptical critic a rightful place and gives historians and philosophers the scope to examine science. Dolly the sheep is here already while the public has committees to decide for them what is good or bad in reproductive medicine and science. This paternalism by interested parties, the state and the industries attached to current research on human, animal and plant 'books of life', a term historically associated with reverence for creation, too, deserves attention, if we are to take autonomous thinking seriously.

Critical debate is required on a serious subject as fundamental as how and what we eat, how we maintain health and how we die. This is why every book that engages bioethics and medical ethics is important. In this one only select themes are written about. But they contribute to considering the basics, namely how medicine and science relate to sociability, that is, the way we form how we live together. The relationship of science to morality and to the sense of where we are heading is crucial. How is money being spent and what does it finance? Do health care systems produce access for all? What are the effects of professional misbehaviour? Are medical men (and it has been gender-specific for a very long time) putting their profession and its corporate interests at the centre of their concerns? What does self-regulation mean and how did it evolve? Finally, how do we decide questions about keeping someone artificially alive and how do we culturally mediate the idea of death in a society dedicated to health, youth, and power?

Many of these chapters deal with German themes, giving readers a rare chance to compare the familiar with historical developments in a sister European nation. More comparative work is desirable, as medical science and its institutions, as well as the structure of the profession, are both different and open to international trends throughout Europe. Particularly interesting are the systems of health care that evolved in each nation and their potential to deliver medical service to the rich and the poor. The allocation of finances and the impact of expenditure on health in different countries take medicine beyond individual rights. It involves politics and government. This too has to be on the agenda and is touched on by many of the contributors to this volume. It shows that influence depends on raising a voice, if not, too, a whole campaign. Ethics have a peculiar nature: they are ancient in their relation to decency and the consideration of others and space age in the problems that society faces.

Modern bioethics has a history in itself and the specific problems it contends with are the ones contemporary society must confront in detail. Since the 1970s Western countries have experienced a surge in bioethics. The institutionalisation of bioethics at universities, the integration of ethicists in hospitals, the important role of ethics committees in the review of biomedical research and health policy-making, and an exploding bioethics literature testify to this remarkable development. This raises the question of the distinctive features of modern bioethics, especially in comparison with traditional medical ethics. One defining influence has been identified in the rapid growth of medical technologies since the 1960s, e.g. chronic haemodialysis, transplantation surgery, modern methods of reanimation, the life-sustaining techniques of intensive care, and in vitro fertilisation.

These and other new technologies have led to ethical problems without direct precedence in the professional medical ethics of the past and thus to a new kind of medical ethics (Jonsen, 1998; Jonsen, 2000). Another interpretation of the move from medical professional ethics to bioethics has highlighted the opening of the hitherto privileged doctor-patient relationship, in its ethical dimensions, by other professional groups, in particular lawyers, moral philosophers, theologians, and sociologists (Rothman, 1991). Finally, the specific social background to the birth of bioethics in the USA in the 1960s has received special attention as a force shaping the current emphasis on rights. The movements for civic rights, equality of ethnic minorities, and women's liberation paved the way for patients' rights and a new approach to ethics in medicine (Reich, 1995; Cooter, 2000; Whong-Barr, 2001).

Tom Beauchamp's and James Childress' four principles of biomedical ethics – beneficence, nonmaleficence, justice, and respect for autonomy – were a landmark in the emerging bioethical discourse (Beauchamp and Childress, 1979, 5th edition 2001). Although all four principles have traditional roots (beneficence and nonmaleficence are already stipulated as proper behaviour for doctors in the Hippocratic Oath), patient autonomy emerged as a new, profound orientation for medical ethics. Thus Ruth Faden, Tom Beauchamp and Nancy King distinguished an old 'beneficence model' of patient information and consent from a modern 'autonomy model' (Faden, Beauchamp and King, 1986). It has since become customary to view modern biomedical ethics as being guided by the principle of patient autonomy. The older medical ethics, by contrast, are usually seen as an expression of the beneficent paternalism of doctors (Jonsen, 2000: 116f). But is it really justified to assume a historical shift from medical paternalism to patient autonomy? Were patients, a hundred years ago, entirely subject to the paternalistic judgements of their doctors? Are patients now, at the start of the twenty first century, fully autonomous in their decisions on matters of their medical treatment?

Taking the 'long view', from the late nineteenth to the late twentieth century, this collection of essays broadly addresses the question of a change from paternalism to autonomy, or 'from medical ethics to bioethics'. Written by authors with academic backgrounds in medicine, history, and philosophy, the chapters of this book unlock the study of biomedical ethics in a variety of ways, each worthy of integration in the overall discussion of an ethical core to medical science and how it developed over time. Collectively, they explore the social, political, legal, economic and clinical dimensions of medical ethics and bioethics with a focus on British and German developments and issues during the past century. Above all they give a sound historical base to what is debated today. History impacts heavily on how problems are seen and solutions sought. Often enough its influence is not analysed or appreciated.

In the first two chapters, Andrew Morrice and Andreas-Holger Maehle discuss the configuration of British and German medical professional ethics around 1900 that structured subsequent developments in ethics. In both countries the notion of 'professional honour' guided the ethical conduct of doctors. Central to this was the relationship to other practitioners rather than to patients. Consultation ethics, appropriate conduct towards colleagues, demarcation from unorthodox or unlicensed healers, and rules against medical advertising were the key issues. The solidarity needed within the medical profession in contracting with mutual aid organisations (Friendly Societies in Britain, Sickness Insurance Schemes in Germany) was also important and contributed to a demarcation against other organised bodies. Doctors in Britain as well as in Germany practised in a highly competitive marketplace. Consequently, much of medical ethics served two objectives: to elevate the public status of doctors above that of other healers and to counteract competition within the profession itself.

With regard to patients the principle of beneficent paternalism, *Salus aegroti suprema lex* ('The welfare of the patient shall be the highest law'), was maintained. However, as both Morrice and Maehle point out, the right of the individual patient to medical confidentiality was challenged, in particular in cases of venereal disease. Obviously contagion raised the obligation of public disclosure and that of the protection of women and children. British and German courts authorised doctors' disclosure of confidential medical information if this served the interests of public health. On another front, however, as Maehle shows for Germany, the law and government action endorsed a person's right to allow or refuse invasive procedures, both in surgical operations and scientific research. Surgery without consent was defined legally as assault and battery; and according to a directive of the Prussian Minister for Religious, Educational and Medical Affairs in 1900, non-therapeutic scientific interventions required informing the subject and obtaining consent. As Morrice makes clear, even when patients were treated paternalistically, they were not powerless. As consumers of medical services made available through insurance schemes, patients could choose their doctor and preclude others.

Several of the themes in these first chapters are expanded in chapters three to five. Lutz Sauerteig traces the social, political, economic, and ethical implications of the German health insurance system since its legal inception in 1883. He analyses the dual role that panel doctors played, deciding which treatment the patient ought to have, while also supplying the recommended treatment. As the gap between expert and lay medical knowledge widened, patients increasingly depended on the judgements doctors made. As insurance schemes began to mediate between doctors and patients, the latter lost their influential role as decision-makers in the direct financing of health care. This decline in medical self-sufficiency, along with economic factors, appears to have supported a paternalistic attitude towards patients. Yet, as Sauerteig's account also demonstrates, the sickness insurance system, jointly financed by employers and employees, contributed to a rapid rise in health costs, especially after the Second World War in the Federal Republic of Germany. This development was certainly driven by consumer (i.e. patient) demand for the best and latest medical treatment and technology. However, the sickness insurance schemes periodically lost control over expenses, initiating re-alignments among the interested parties and changing the way costs were met. Sauerteig highlights the public perception of a 'cost explosion' in health care in the 1970s, coinciding with a world-wide economic recession, which decreased the income of the insurance schemes. 'Cost containment' and calls to fight a 'claim mentality' entered the arena of health politics. Sauerteig shows changes in the ethical priorities of German health care policy resulting from these economic developments. Before 1970, the idea of social equality was important, expressed by efforts to widen patients' access to professional medical services. Afterwards a discourse on scarce resources and their allocation supplanted the ideal of a fair distribution. Indeed, problems of resource allocation have now become a common feature of bioethical debates. In recent years, financial burdens, such as higher insurance premiums and personal cost sharing, have come to rest on the ability to pay of the patient, undermining principles of social solidarity. In the light of Sauerteig's analysis we may ask: is patient autonomy in health care decision-making now seriously compromised by a 'privatisation' of health costs?

In chapter four, Cay-Rüdiger Prüll and Marianne Sinn analyse the issue of informed consent, exploring in detail patients' and relatives' reactions to autopsy as part of medical research and to new surgical methods practised on patients in early twentieth century Germany. Between 1900 and before the National Socialist dictatorship of 1933, patients increasingly demanded a voice in decisions on medical treatment, a phenomenon that was accompanied by a growing public interest in 'alternative' therapies, especially in naturopathy. Public resistance against post-mortem examinations went hand in hand with demands for more information about and better means of obtaining consent as surgical operations reached a new zenith. The response of doctors, however, depended – as Prüll and Sinn show – on how the profession might be disciplined. Pathologists, threatened by public criticism of one of their basic scientific tools, the autopsy, tried to emphasise the medical usefulness of post-mortem examinations. They preferred to circumvent specific consent-seeking from the deceased's relatives. Surgeons, on the other hand, were buoyed by their success in areas like abdominal surgery, and could face the issue of consent from a position of strength. While courts of law continued to hold the view that operations without consent (or even against the patient's will) were culpable as battery, surgeons insisted, paternalistically, that patients were incapable of fully understanding the implications of medical interventions and therefore had to be 'guided'. In fact, 'implied consent' often relied on rather basic information about an operation, and this was deemed sufficient even from a legal point of view. Prüll's and Sinn's chapter demonstrates that there was no uniform move in medicine honouring patient autonomy and that the different contexts of each medical speciality changed the ethical approach.

In chapter five, Ulrich Tröhler expands the issue of informed consent to cover human experimentation, which became a key topic in bioethical debates and codifications. He observes that while in the eighteenth century consent was not an issue, the demand that experimentation should benefit the patient was often voiced. Only after abuses in human trials in the second half of the nineteenth century was information and consent seen as a requirement. The Prussian Minister for Religious, Educational and Medical Affairs made it mandatory in 1900. Although guidelines on human experimentation were issued in expanded form in 1931 by the German Ministry of the Interior, these did not prevent the atrocious human experiments in the concentration camps of Nazi Germany carried out by doctors. In the Nuremberg Trials of 1946/1947 these criminal abuses were revealed to the world and generated the *Nuremberg Code* (1947). As Tröhler points out, this was the first document in international public law that included the requirement for voluntary and informed consent by human subjects and anchored their right to abandon an experiment at any time. In the subsequent numerous national and international codes regarding clinical trials, Tröhler distinguishes two types. The older type, exemplified by the *Declaration of Helsinki* of 1964, was still grounded in paternalistic views and, despite the shock of Nuremberg, still tried to ensure that human experiments continue, albeit by protecting human subjects. The second, more contemporary type adopted the language of 'rights' originating in the late 1960s and, more recently, the concept of human dignity. An example for this is the *Convention on Human Rights and Biomedicine* (1996) of the Council of Europe. While Tröhler's contribution, as a whole, portrays a move from paternalism to autonomy, it also shows the limitation of bioethics codes. Whether they are produced by non-governmental organisations, such as the World Medical Association, or by inter-governmental organisations, such as the Council of Europe, these codes provide little more than a generalised basis for legal liability or at best a guideline to better medical practice. For a real impact they must engage with current debates and have practical applications.

Another major concern in recent bioethical debates, where the tension between paternalism and autonomy surfaced, is medical treatment at the end of life. Chapters six and seven of this volume engage this problem. Bryan Jennett looks into the challenges medical paternalistic decision-making raised for life-saving and life-sustaining treatments. In the USA court decisions supported the withdrawal of treatment for terminally ill patients in the mid-1970s and encouraged the use of advance directives ('living wills') or joint decision-making between doctors and relatives of patients, if these could no longer express their wishes. In Britain, doctors' paternalism in end of life decisions remained largely unchallenged until well into the 1980s. The case of a teenage Persistent Vegetative State (PVS) patient, Tony Bland, in 1993, changed this situation when judgements in the High Court, Appeal Court and the House of Lords declared the removal of his feeding tube legal. Since then, the Official Solicitor has stated the need for formal court approval in such withdrawals, and relevant guidelines were issued by the British Medical Association and the Royal College of Physicians of London. In his discussion of the decision to withdraw or withhold life-sustaining treatments, Jennett argues for an increased use of advance directives and consultation with relatives to make sure of the likely wishes of mentally incompetent patients. He also discusses the problem of prognostic uncertainty and the role of clinical judgements that cannot rely on absolute medical knowledge, pointing out that in the UK (in contrast to the USA) only a doctor can decide to end treatment. Jennett's chapter illustrates recent efforts to guarantee patient autonomy in end of life decisions, but also draws our attention to the practical limitations of such efforts.

A very fundamental question, whether patient autonomy gives a person the right to make a doctor comply with a request to be helped to die, is examined by Susan Lowe. She argues that neither voluntary active euthanasia nor physician-assisted suicide is justified in the name of patient autonomy. Two main reasons are presented for this. Firstly, while suicide is legally permissible (in Britain), this does not imply a *right* to commit suicide with which others must comply. A doctor refusing a patient's request for assisted suicide or for active euthanasia does not violate patient autonomy. Secondly, demanding compliance with such a request against a physician's will violates that physician's autonomy. Lowe's chapter makes it clear that there are not only practical limitations to the exercise of patient autonomy, but also faults in the logic of how patient autonomy is argued.

David Cooper invites us to rethink basic problems on the cutting edge of bioscience, particularly genetic engineering. He reminds us that public perceptions of medical science were never those of an uncritical acclaim.

Taking *Frankenstein*, Mary Shelley's perennial best-seller, as his cue, he shows that 'fear and revulsion' in the face of interventionist biology is not uncommon, then or now. This has little to do with an alleged public hostility to science, nor does it mean the general public underestimates scientific benefit. Nor does it have to do with the alleged ignorance of common folk about complex scientific research. Patient autonomy, after all, is based on the concept that treatments and their consequences can be understood. In fact, as Cooper nicely points out, those sceptical of science have had to endure a large measure of abuse about backwardness and obstructing the better future of mankind. Genetic modification of crops, creation of transgenic animals for organ transplantation, cloning, and germ-line therapy is not by its scientific nature a moral good. It raises considerable ghosts because utilitarianism has never been nor is it the sole arbiter of moral thinking. Making judgements about what is beneficial in a larger sense includes a reverence for life, and not just that of human beings. As Cooper states, *Frankenstein* is not about the monster, but about human intervention in the generative process of (human) life. Scientific intervention in the life process is of another order than most manipulation. It calls forth misgivings about 'the human endeavour to mock the stupendous mechanism of the Creator' as Mary Shelley put it. Having attained the status of an 'autonomous industry' it begs the question whether all sanction its workings for whatever benefit. People's immediate fear of technological intervention relates to a vivid sense of powerlessness when faced with closed professional worlds, a lack of inclusion and openness the word paternalism only hints at.

The recent public outcry over the collection of pathological specimens from prematurely born babies without gaining the explicit consent of relatives (House of Commons, 2001) alerts us once more to the importance of human feelings. They legitimately express the respect no scientific rationale should lightly violate. Moral feelings and values are not exhausted by tensions between rights and duties. Autonomy and paternalism name just one variant in the debate. But we want it to be a strong and rambunctious one. In the light of this final chapter we may raise the question: Will the New Genetics, rather than enhancing people's choice (e.g. through pre-natal testing), actually pose a threat to patient autonomy in the twenty first century?

Readers of this volume will form their own opinion on the shift from paternalism to autonomy in twentieth century medicine. The Editors believe that these two concepts provide a useful heuristic tool for the critical study of ethics in medicine, past and present. The paternalism of old, universally criticised today, enfolds within it such positive qualities as taking responsibility for professional conduct and having proper qualifications for treating patients. A 'manly' virtue, it evokes old-fashioned duty and caring. The call to autonomy has dominated late twentieth century debates on rights. But it too has pitfalls. In particular it threatens actions in isolation, be they corporate, as in a medical science oriented towards profit and without accountability to the general good or be they personal. An autonomy of self-righteousness destroys the ability to negotiate and work together.

As far as the evidence of this volume goes paternalistic medical practices and patient autonomy had an uneasy relationship by the beginning of the twentieth century. A hundred years later, full autonomy in decisions on medical treatment is still subject to numerous caveats. Current developments in the economics of health care and in biomedical science may even counteract efforts to enhance patient autonomy. Ethics in medicine has been highly dependent on social contexts. The medical profession, courts of law, health insurance schemes, governmental organisations, and the biotechnological industry continue to be key influences. Close attention to how this interplay develops and how historical patterns have shaped responses in the many contexts of our sociability – not only its underlying philosophical principles – will decide the bioethics of the future.

References

Beauchamp, T.L. and Childress, J.F. (2001), *Principles of Biomedical Ethics*, 5th edition, Oxford University Press, Oxford.

Cooter, R. (2000): 'The Ethical Body', in idem and Pickstone, J. (eds), *Medicine in the Twentieth Century*, Harwood Academic Publishers, Amsterdam, pp. 451-68.

Faden, R.R., Beauchamp, T.L. and King, N.M.P. (1986), *A History and Theory of Informed Consent*, Oxford University Press, New York.

House of Commons (2001), *The Royal Liverpool Children's Inquiry: Summary & Recommendations*, The Stationery Office Limited, United Kingdom.

Jonsen, A.R. (1998), *The Birth of Bioethics*, Oxford University Press, New York.

Jonsen, A.R. (2000), *A Short History of Medical Ethics*, Oxford University Press, New York.

Reich, W.T. (1995), 'Introduction', in idem (ed.), *Encyclopedia of Bioethics*, revised edition, vol. 1, Macmillan, New York, pp. xix–xxxii.

Rothman, D.J. (1991), *Strangers at the Bedside: A History of How Law and Bioethics Transformed Medical Decision Making*, Basic Books, New York.

Whong-Barr, M. (2001), 'Medical Ethics in Historical Contexts', *Medicine, Healthcare and Philosophy*, vol. 4, pp. 233–5.

Chapter 1

'Honour and Interests': Medical Ethics and the British Medical Association

Andrew A.G. Morrice

In this chapter, which is based on a much larger body of research (Morrice, 1999), I shall examine the nature and content of medical ethics in Britain during the late nineteenth and early twentieth centuries. The focus will be primarily on the British Medical Association (BMA), and the work of its Central Ethical Committee (CEC) particularly between 1902 and 1948. I shall also discuss the basic historiographical problems raised in writing a history of medical ethics, as well as making more general comments on the light this study may shed on the social function of medical ethics.

Introduction

For many readers the first question will be, 'why the BMA?'. The BMA's ethical work drew on written and unwritten rules dating back over the previous hundred years, and related closely to general notions of inclusion and exclusion embodied by the ancient medical corporations. The work of the ethical committees was seen as integral to the Association's aim to 'maintain the honour and interests of the medical profession', itself integral to the professionalisation of medicine in Britain. The Association attempted to influence and work with the statutory body overseeing medical practitioners, the General Medical Council (GMC), and with professional defence bodies, particularly the Medical Defence Union (MDU). By the early twentieth century the BMA represented the majority of doctors practising within the British Isles, and many of those practising in the rest of the British Empire, and had amongst its membership practitioners of every kind and status. In contrast to the other organisations dealing with doctors' conduct, it had to debate and decide its policies with reference to its members. Both the 'codes' of 'medical ethics' published by British authors between the BMA's founding in 1832 (as the Provincial Medical and Surgical Association) and the Second World War were produced by men with active and important links with the BMA, indeed the second was written by a chairman of the CEC (de Styrap, 1878, 1886, 1890, 1895; Saundby, 1902, 1907). The Association was described in 1933 as having 'gone from strength to strength' and as perhaps the most successful professional body of its times, 'credited with powers it d[id] not in fact possess'.[1] These factors, and the survival of extensive archive materials in accessible form, make the BMA's ethical work perhaps the single most informative window onto the relationship between early twentieth century medical ethics and the wider context in which the profession was working.

After the Second World War the CEC finally produced a code of medical ethics (BMA, 1949), a task that had been first proposed 116 years previously. It can be read as a summary of their work in the previous 45 years, which was itself based on ideas dating back a further century. Yet it was at this historical moment that two documents – the *Nuremberg Code* (1947)[2] and the *Declaration of Geneva* (1948)[3] – both arising from the aftermath of the Second World War, prefigured a quite different kind of medical ethics. This newer, and to modern eyes more recognisable medical ethics had, by the 1970s and 1980s, almost entirely eclipsed the kind of issues on which the BMA had laboured so long.

An initial survey of 'medical ethics' between the turn of the nineteenth century – when Thomas Percival chose the phrase to title his codification of professional behaviour (Percival, 1803) – and the mid-twentieth century would show that it was concerned largely with doctors' behaviour towards other practitioners. Thus the arrangements for conferring over cases without threatening the original doctor (consultation ethics) featured strongly, along with matters of propriety, mutual respect for medical brethren and the avoidance of disputes. Later in the nineteenth century, relationships with unorthodox healers and the strict avoidance of anything that might be construed as advertising had become prominent issues (de Styrap, 1878). By the turn of the century the relationship of doctors to lay organised medical services, whether mutual self-help organisations or commercial enterprises, joined the list of major concerns. The most familiar looking issue in this canon of medical ethical writings is the topic of confidentiality, or secrecy, as it was sometimes known. Writers on medical ethics were doctors, and the disciplinary and ethical organisations of the profession involved no lay people.[4] Moreover, the application of moral principles was limited to general appeals to notions such as the 'Golden Rule' ('do as you would be done by') which was generally invoked in relation to other orthodox practitioners alone.[5]

Critics of the medical profession have long found much to deplore in the ethics produced 'by and for' the profession which characterise the period up to the Second World War (Leake, 1927; Roberts, 1937; Waddington, 1975; Freidson, 1975; Berlant, 1975). Such critics, along with leaders of medical opinion in the post-war years, have tended to re-categorise the earlier professional ethics as 'mere etiquette' and have made much of the way that professional ethics underpinned the power and prosperity of the medical profession.[6]

Modern medical ethics involves many experts in moral philosophy, and therefore the detailed deliberation of a large number of philosophical ideas, and addresses issues primarily of concern to patients. Its rise has been part of the social trend to limit and modulate the power of the medical profession.[7] The behaviour of doctors toward each other now receives little attention aside from injunctions to report wrong-doing or poor performance promptly (Doyal and Gillon, 1998). The issues discussed are usually matters involving the extremes of medical intervention and the fundamental events of life: reproduction and terminal illness, and a set of ethical principles, most prominently the right of autonomy. Perhaps most significantly, medical ethics has become an academic subject in its own right, with a rapidly burgeoning literature, 'so copious that any physician wanting to keep abreast of [it] would have to abandon the practice of medicine',[8] and it is now possible to pursue a career in medical ethics (Jonsen, 1990 and 1998).

Thus the content and form of the writings, ideas and adjudicatory structures denoted by 'medical ethics' have shifted radically during the 200 years in which the phrase has been used in the English speaking world. Had Percival listened to some of his correspondents, who urged him to call the book 'medical jurisprudence',[9] or had he the clairvoyant capacity to respond to his critics in the twentieth century, and had published *Medical Etiquette*, the historian's task would be subtly but significantly different. Since the medical professions of both the United Kingdom and the United States of America looked to Percival as the originator of their medical ethics up to the middle years of the twentieth century, this is no minor consideration.[10] As it is, Percival was swayed by the notion that his work was based on moral ideas rather than legalistic ones, and recognised the distinction between the etiquette he proposed, and the ethics underlying it. He felt, like his friend and correspondent, Thomas Gisbourne, who had published his *Enquiry into the Duties of Men in the Higher and Middle Classes of Society* in 1794, that there was a profound moral dimension to the advice they offered. They were also drawing on ideas of gentility and gentlemanly conduct, which became a core aspirational, self-defining ideal in British medicine, particularly during the late nineteenth and early twentieth centuries (Peterson, 1984; Lawrence, 1985).

Thus at the heart of any endeavour to write the history of medical ethics lies an intellectual and historiographical problem. This dilemma can be viewed as a semantic problem generated by the apparently changing content of 'medical ethics'. It could be viewed as a conflict of methodology between social historians, who have recently dominated the history of medicine, and those primarily concerned with the current discipline and philosophical basis of medical ethics. It may be unrealistic to expect moral philosophers engaged in establishing their role as adjudicators of medical rights and wrongs to look for moral content in a professional ethic that can so easily be dismissed as self-serving. For social historians the problem can be viewed as a choice between the argument that the content of medical ethics has simply changed as the social context and day to day work of doctors has changed, or that beneath this apparent shift lies a more profound continuity of social function. Those interested in the subject should, perhaps, take their lead from Mary Warnock, who writing an account of moral philosophy in the twentieth century stated she 'ha[d] not, I hope had any preconceived idea of what ethics is' (Warnock, 1960). For this particular social historian, the most useful guidance as to what, in social historical terms, medical ethics might be, has been articulated by Roger French. He says:

> The current interest in medical ethics is an interest in ethical problems. It might seem unproblematic that medical ethics have a history, and that these problems can be studied in the past. We might for example take the problem of abortion and look at it historically. But ... such a history would be the history of a practice, not of an ethical problem. ... modern medical ethics derives from the particular nature of modern medicine and the society in which it exists. So a history of medical ethics is a history of medicine and of society and of the problems that looked ethical to them, but not necessarily to us. Looked at in this way it soon becomes clear that ethics have a function, for the group that practices them, other than the internal, explicit injunctions that are normally seen as 'ethical' in some abstract way. ... Ethics comprise a system of rules that not only characterises the group but which in directing the behaviour of the group contributes to its success. (French, 1993:72)

Doctors operate within a set of forces that shape their behaviour. Some of these are very fundamentally cultural, some legal, some arising from the statutory functions of medical organisations, some contractual, some frankly economic, whilst some derive from ideals of what a doctor should be, generated not only by society but by the profession itself. My research has been based not on tracing back current ethical concerns in history, but on looking at the organisational structures and procedures described by contemporaries using the terms 'medical ethics' and 'professional conduct', and literature related to them. However, since this overlaps with legal questions relating to medical practice and with statutory professional discipline, I have simultaneously used a second definition or frame. This defines medical ethics as *setting standards for, and adjudicating between right and wrong medical behaviour, where this is not defined by law*. I shall describe the forms of this advice and adjudication, and the assumptions and ideals underpinning it. Taking my cue from French I have attempted to describe the way these ideas and procedures structured and characterised the medical profession, and how, and to what extent they 'contributed to its success'. It seems to this author *naive* to suppose that in an era characterised by the sometimes belligerent defence of group interests in society (Harris, 1993), and structured around notions of class and respectability (Thompson, 1988), that doctors would not organise and promote norms of behaviour that secured and promoted their place in society. In the light of Perkin's work, it would be strange to assume that medical ethics would function to undermine a profession's socially sanctioned role (Perkin, 1989). In fact those involved in the BMA's ethical work frequently and explicitly – if in modern eyes rather simplistically – linked the profession's own interests with 'the public interest', and recognised that the ideal was a situation in which the needs of doctors and patients could be harmonised.[11]

Medical Ethics, Professionalisation and the BMA in the Nineteenth Century

Prior to the Medical Act of 1858 the existence of a medical profession in Britain was a medico-political ambition rather than a statutory reality. Up until this point the behaviour of medical men might be influenced by writers such as Percival, John Gregory, or Gisbourne, and they might be subject to local codes and conventions, but formal disciplinary power was only wielded by the medical Colleges and corporations. These organisations grew out of the University and guild systems of early modern Europe, initially in the Italian city states, and served to promote the status and interests of their members. This was accomplished by two basic functions, which are echoed in the workings of most medical organisations: admission by qualification, and expulsion for unsuitable conduct. The disciplinary functions and strictures of the Royal Colleges prefigured and were probably the template for much of the form, tone and content of later disciplinary and guidance functions of the newer medical organisations of the nineteenth century. For example the Royal College of Physicians of London, founded in 1518, was an elite organisation that admitted to Fellowship only medical graduates of Oxford, Cambridge and the continental universities. The College Fellows through their *Comitia* and Censor's Board exercised disciplinary powers over lower grades of practitioners, originally London's apothecaries, and subsequently their Members and Licentiates (Clark, 1964). Essentially any member of the College guilty of 'crime or immorality', or whose behaviour was found by the Censors to be 'dishonourable or unprofessional', could be voted out of the College by a majority of the Fellows in *Comitia*.[12] Whilst some behaviours were covered by the Colleges' Bye-laws and Resolutions, specific grounds for expulsion were only partially codified.

The founder of the BMA, Charles Hastings, alluded in 1833 to the lack of a 'well digested code' to augment the few specific injunctions of the Colleges and the ideals adumbrated in published texts, and felt that his new Association should be able to produce this more detailed and practical guidance.[13] However, it took the Association 70 years to establish productive ethical committees, and far longer to publish an ethical code. This was an embarrassment in contrast to the American Medical Association (AMA) which was almost defined by an ethical code agreed within a year of its inauguration in 1846. The AMA code was broadly based on Percival's *Medical Ethics* (Baker, 1995). The BMA appointed two ethics committees in the 1850s, which were instructed to produce similar codes (Horner, 1995). They failed to meet, agree or report, although one member of the second failed committee, Jukes de Styrap, went on to compile an updated version of Percival in 1878. This *Code of Medical Ethics* went through several editions and enlargements, down to 1895. Significantly de Styrap's work grew out of his involvement in one of many local 'medico-ethical societies', many of which became branches of the BMA, and like Percival he drew on local codes and the opinions of correspondents. De Styrap also published several editions of the fees and other charges agreed in his own locality, Shropshire.[14] The activities of such local groups appear to have constituted a significant proportion of medical ethical activity in Britain up to the 1890s, but it is hard to gauge precisely how extensive or general these activities were. In his *Code* de Styrap's main concern was to promote harmony and prevent disputes within the profession, and to denounce unqualified and unorthodox healers, and 'quacks'. The main departure from Percival related to the shift in medical politics following the drive for medical reform, and the Quackery Scare of the 1830s and 1840s that had been integral to it (Bynum and Porter, 1987). For the first time there was a statutory definition of the profession, and this firm boundary was jealously patrolled in the form of injunctions not to associate with unregistered practitioners of any kind.[15] A further difference lay in the increased prominence of injunctions against advertising in any conceivable form.[16] This reflected the need to differentiate doctors from quacks and tradesmen.

Thus the first code produced in association with (but not by) the BMA was profoundly influenced by existence of the statutory *Register of Medical Practitioners*, for which the Association had fought long and hard. The 1858 Medical Act had not provided all that the BMA and other medical politicians had campaigned for, and the BMA continued to push for a more energetic, representative and medico-politically relevant General Medical Council (GMC) well into the twentieth century. The Act's most important 'omission' was that unregistered practitioners were not debarred from practising medicine, only from claiming to be registered. Entry to the *Register*, like that to the Colleges, was by qualification, and expulsion was the punishment for 'crimes and felonies' or poor conduct, defined by the Act as 'infamous conduct in a professional respect'. The methods by which this disciplinary function were to be carried out had not been defined by the Act, and the fledgling Council had to extemporise a procedure when faced with its first serious complaint.

The Council opted for a kind of pseudo-judicial process, with the Council members acting as judge and jury (Smith, 1993). The practitioner concerned appealed against the Council. His complaint that the procedures were unfair has been repeated frequently down to the present day (Smith, 1994).

There was initially absolutely no definition of 'infamous conduct in a professional respect', and judicial responses to appeals against the Council effectively left it free to decide what 'infamous conduct' meant in any case (Harper, 1912). Later the BMA and MDU persuaded the GMC to produce a limited set of warning injunctions. These pronouncements, which came largely in the late 1890s and 1900s, all had a bearing on the medico-political crisis of the time, which is discussed below. Removal from the *Register* had limited effects on a practitioner however. Those who had a university doctorate could still describe themselves as Doctor, and any citizen could practise medicine so long as they did not claim to be registered. During the twentieth century new laws and statutes allowed only registered practitioners to do certain things, and occasionally prevented the unregistered from doing others. Aside from this increasing number of legal obstacles, the only hindrance to the working lives of unregistered medical practitioners was the hostility of the registered profession, and particularly the ethical rules that forbade the registered from working with the unregistered (Minty, 1932).

By the last third of the nineteenth century, a number of pressures combined to produce a medico-political and medico-ethical agenda (and these two terms were used together) which resulted eventually in the reform of the BMA[17] and the establishment of its ethical committee system in 1902. The problems were mainly caused by medical overcrowding, by the availability of free care in the outpatient departments of voluntary hospitals ('hospital abuse') and the control of medical work by various non-medical authorities. Of these latter, the most contentious were the working and artisan class mutual aid arrangements, usually known as Friendly Societies. These societies were eventually organised under government control to provide National Insurance in 1911. By the 1890s the profession and these mutual aid organisations were engaged in a continual struggle for control and remuneration (Green, 1985). The key objections of doctors were low capitation rates, excessive workloads, lack of medical control of these services, and the lack of wage limits to exclude subscribers who could afford private fees. Both doctors and Friendly Societies became more organised and intolerant, and disputes between them often resulted in doctors' attempting to deprive the societies of medical manpower. In these disputes medical solidarity was the key to success, and doctors breaking local boycotts of contract practice were often ostracised by their fellows.[18]

Many ordinary doctors felt that their professional bodies had failed to represent them adequately, and this dissatisfaction had lasting effects on both the BMA and the GMC. Attempts were made to give the dissatisfied a more prominent voice in BMA affairs. For two years an 'ethical section' was held at the BMA's Annual Meeting, and the General Practice committee was given a role to consider conduct issues. These measures had no discernible effect, and by 1900 there was a real danger that the BMA would be superseded by a novel organisation of Medical Guilds. The BMA responded to this crisis by completely overhauling its organisation and structure, and became for the first time in its history a truly democratic and dynamic body (Cox, 1950; Bartrip, 1996). During the 1890s the GMC found itself under pressure from a newer and at that time more energetic body than the BMA, the MDU. This pressure was applied not only through lobbying but also in bringing cases before the Council. The GMC was eventually induced to pronounce formally and publicly that certain behaviours constituted 'infamous conduct in a professional respect', and published this information as occasional 'warning notices' in the medical press. These behaviours were: 'covering unqualified assistants' (enabling an unregistered practitioner to work as if he were registered or to do the work of a registered practitioner) and 'advertising and canvassing'. These activities characterised, but were by no means limited to contract practice arrangements organised by Friendly Societies. In fact in many ways they acted to make contract practice less profitable and tenable for many of the doctors engaged in it, thus exacerbating the problem. According to Robert Nye, similar pressures appear to have been important in French doctors' increasing interest in *déontologie* at the same period (Nye, 1995). Andreas-Holger Maehle's chapter in this volume also points out that overcrowding of the profession and relationships with insurance organisations were key to pressures to establish the 'courts of honour' system in Prussia during the late 1890s.

The New BMA and its Ethics Committees

The new BMA was a federal structure with local Divisions (drawn up so as to render meetings easily attended by all their members) and above them regional Branches, a central secretariat, and a Council. Most importantly, all BMA policy was decided by the Annual Representative Meeting (ARM), a vibrant and at times unpredictable body made up of Representatives of the local Divisions.[19] It would be hard to overstate the importance or success of this constitution, which made the Association vastly more powerful and influential. Each Division was intended, amongst other things, to act as a 'Court of Honour' (almost certainly a borrowing from the German term) to decide ethical issues and disputes.[20] The Central Ethical Committee was set up to advise these local groups, and to take on matters that neither the Division nor its Branch could decide. The Committee was also responsible for presenting cases for expulsion from the Association, under the general rubric of conduct 'contrary' or 'detrimental to the honour and interests of the medical profession' – a rubric that echoed both the Royal Colleges' and the GMC's disciplinary powers. The Committee was also charged with drawing up reports or sets of general rules on particularly difficult issues.[21] This was the first time ordinary doctors in Britain had a national body to examine questions of conduct without resorting to the extremes of hearings before the Colleges or GMC, it was also the first time a national body had been available to settle disputes between individual practitioners.

The Central Ethical Committee soon became very busy indeed, and by 1906 it was decided to establish a standing sub-committee of members living near London.[22] (Its immediate predecessor committee, the General Practice and Ethical Committee, under the same chairman, had become steadily less and less active between 1895 and 1902, probably because the BMA structure as it then was gave its deliberations no clear force or status.) There was a meeting of either the main or subcommittee once a month, except during the summer, and they dealt with large numbers of inquiries, cases and contentious reports. Its ordinary members were for the most part modestly well qualified, typically practitioners of general medicine, many with substantial surgical practice, and some with minor, local posts as specialists. They were usually in their 50s or 60s, having 20–30 years experience of practice when they joined the Committee, most of them lived in the south-east of England. They were probably representative of the kind of doctor with the time, security and logistical opportunity for this work. The Committee was chaired by slightly more eminent men, with a greater preponderance of 'consultants', particularly in the period before the First World War. Its first chairman, Professor Robert Saundby, went on to produce a key text, *Medical Ethics*, based on his experience in the BMA, and became a member of the GMC.

By the interwar years, a core of long-serving members had been built up. One member, Reginald Langdon Langdon-Down attended around 300 ethical committee meetings between 1907 and 1939, and a small group of 30 practitioners made up 50 per cent of the attendances at these committee meetings.[23] Significantly the secretariat staff assigned to the ethics committees were often destined for higher office. Alfred Cox and George Anderson both became (Medical) Secretaries of the BMA, and James Neal and Robert Forbes both became Secretaries of the Medical Defence Union.

Sadly it is impossible to ascertain quite how the committee functioned when it met. Each meeting had large numbers of memoranda and other documentation to consider, and often met practitioners faced with expulsion from the BMA, or doctors involved in specific disputes. It seems reasonable to assume that the chairman and the secretary had significant power and discretion. Except when asked to do so by the ARM, the committee displayed a marked distaste for answering general or 'hypothetical' questions, and resisted publication of a digest of their decisions on the grounds that the minutes and decisions as recorded omitted 'important points ... in the discussion which had a material bearing on the conclusions'.[24] Their thinking is therefore more apparent in the deliberation of the general issues that the ARM or Council asked them to consider.

Professional Medical Ethics in the Early Twentieth Century

What were the concerns in medical ethics at this time? It is clear that at any one time there were only one or two major issues on which the committee was expected to provide general guidance, and that these issues relate to each other in particularly interesting ways. It is also clear that these were major themes and concerns in the published literature. My discussion therefore will focus on these major themes. But before starting on these, it is worth pointing out the low profile of many apparently 'ethical' issues. In my reading of the CEC and subcommittee minutes between 1902 and 1939 I have found no enquiry or statement about experimental ethics,[25] no requests about patient consent to treatment,[26] one enquiry about the effects of palliative drugs and euthanasia, one about abortion,[27] and enquiries about contraception only in terms of local competition and the setting up of advertised clinics.[28] This is not to say that doctors were not concerned with the moral dimensions of these areas of medical practice. They simply do not appear to have thought of them as *medical* ethics – they were, rather, part of wider religious and social beliefs, and perhaps more importantly still, matters for general rather than medical adjudication.[29]

Solidarity and Disputes

The first great tasks of the CEC were to draw up rules to support the fight against lay controlled contract practice, and to guide the ethical work of the Divisions. As was noted earlier, the failure of some doctors to comply with local boycotts in disputes over pay and conditions was seen to be a key problem. Most texts on conduct stressed that a doctor coming to an area should call on his local colleagues to establish his place in local medical society, and that doctors should avoid acting to the detriment of one another. The BMA rules governing local groups, and on appointments in particular, radically extended the way in which these connections impinged on individual doctors. Drawing on local resolutions and rules which they had been sent for approval or comment, themselves based on more general notions of collegiate obligation, the CEC produced model rules for Divisions and Branches. These, amongst other things, defined and placed on members a set of duties to co-operate in the withdrawal of medical services from posts which the Division had declared 'detrimental to the honour and interests of the medical profession'.

From 1903 a 'Warning Notice' was introduced to the members' *Supplement* to the *British Medical Journal*, which listed areas in which disputed posts were vacant, with an instruction to contact the Honorary Secretary of the local Division before applying.[30] A doctor failing in this specific regard, or who took up a disputed post, or whose behaviour was in any other way objectionable, could be called to a meeting of the local Division who could declare his conduct 'contrary to the honour and interests of the medical profession'. It could then become the duty of local doctors to refuse to meet or consult with this doctor. Anyone doing so was liable themselves to be found guilty of the same offence.[31] Many doctors holding disputed posts were forced to resign, and many were expelled from the Association.

Interestingly, in other respects the model rules promoted the friendly and swift resolution of other kinds of disputes between doctors. Great emphasis was placed on avoiding public rancour, and a gentlemanly agreement to differ or to change offensive behaviour. It is important to note that there were procedural rules only, there was no code of specific ethical offences to go with them. The local doctors were thought to be already competent to decide the issues at hand, or if in doubt, could refer on to the more senior, more experienced, but essentially similar Central Ethical Committee.[32] All of these concerns, rules, and structures were a sophistication and formalisation of ideas and forms of behaviour that had existed all over Britain during the late nineteenth century, but which had far more power in the context of a national organisation.

It is worth looking at the phrase 'honour and interests' here. It was obviously in doctors' interests to gain control of the Friendly Society and other provident medical services, in order to raise their rates of pay, and exclude subscribers who could afford private fees. There was also a profound social dishonour felt by many doctors in being employed by committees of men who were putatively their social inferiors. Once a local BMA Division, with the backing of the central organisation, declared that a certain job was 'contrary to the honour and interests of the medical profession', doctors who broke ranks were reclassified as 'outsiders' and excluded from professional and social intercourse. The machinery of the local rules and ethical committees, and particularly the 'Warning Notice' – which was legally appraised as constituting a 'black list'[33] – came into question in a series of cases between 1908 and 1915, but the arrangements were only altered slightly. These local campaigns could become unpleasant, and one group of ostracised doctors from Coventry took the BMA to court and won damages of libel, slander, and conspiracy in 1918.[34] This dishonoured the Association publicly, and the boycott scheme itself was once again modified. However the basic structure and the ideas underpinning it remained in place until 1948.

Consultation

As we have seen, ostracised doctors were excluded from consultation with colleagues, and the general topic of consultation was the next important area on which the CEC was asked to report. Medical practitioners of all sorts were reliant on the help and advice of more expert or experienced doctors to deal with difficult cases. However, since the incoming, usually senior practitioners could easily supersede the attending practitioner, either at the invitation of the patient or her friends, or through deliberate manoeuvring, the doctor in attendance had to balance two risks. He could risk losing the patient for want of advice or assistance, and further patients through loss of reputation, or risk losing the patient to the other doctor, and with her further patients.[35] The issue of consultation ethics seems to have been, rather like the contract practice issue, a long-term grumbling problem,[36] which the BMA was in a position to solve in a new kind of way.

The Committee produced detailed guidance, but this was based on previous codes and was presented as an expression of 'best traditions' of medical behaviour. Their reworking of these 'traditional' ideas took place however within an organisation that aimed to unite and represent all kinds of doctors, and the Committee had to draft their rules in dialogue with the Council and ARM. In fact it was never possible to produce a set of rules agreed in all details by the ARM, a fact that shows how contentious the detail of this issue was, and which also reflected widely divergent conditions and customary arrangements in different parts of the country.[37] Another bone of contention with some Council members was the refusal to create a definite category distinction between what we now term Consultants (specialists) and General Practitioners (or family doctors).[38] The term consultant in this context meant any doctor called in, by the doctor who usually cared for the patient, who was termed the attending practitioner. The assertion that there was a continuum of types of practice certainly appears true of the professional lives of CEC members.

The place of consultation ethics in Percival's *Medical Ethics* and its derivatives was the grounds for the charge that these works were simply etiquette rather than ethics proper. In fact the CEC's report on consultation recognised, just as Percival had, that the specific recommended behaviour constituted an etiquette which was founded on ethical principles. The report stated these to be that it was the attending practitioner's *duty* to avail his patient of the best available advice, and that the incoming consultant doctor was in a position of *trust* with regard to his colleague. The purpose of the rules was to encourage appropriate consulting – felt to be in the patient's interests – by protecting the position of the attending practitioner. It proved to be very difficult to prevent patients changing doctors without breaking these conventions, since patients were reluctant to go and formally break off their previous medical relationship before engaging another attending practitioner. Patients were often keen to approach specialists or consultants themselves without reference to their usual doctor. Unsurprisingly, consultation etiquette was unpopular with the public, and patients often complained that their choice was being restricted, and suspected they were being exploited (Sprigge, 1905 and 1928). The BMA was acutely aware of this, and during the 1930s the CEC and BMA secretariat struggled to defend the rights of patients in choice of doctor, as well those of the attending practitioner. However, many Representatives at the ARMs were opposed to any loosening of these rules, which were seen by many as important traditions, founded in basic social obligations, and also important to protect the health of patients. The relationship between the patient and their attending practitioner seems almost to have had 'matrimonial' status (Brackenbury, 1935). The traditional English marriage service uses the phrase 'those whom God has joined together let no man put asunder'. It seems that medical ethics was designed to ensure that those whom the mysterious forces of the medical marketplace had joined together, no medical man was to put asunder, at least not without the knowledge and acquiescence of the attending doctor.

The importance of referring patients on to the best expert consultants was also apparent in an important but little discussed problem, that of dichotomy or fee splitting. This practice appears to have been carried on amongst specific groups of practitioners, so that some would declare it rife, and others state they had never once come across the problem. Fee splitting was usually presented as a vice of foreigners, a rarity in Britain, and roundly condemned whenever it was discussed.

Other Practitioners

By the last third of the nineteenth century medical ethical texts advised the exclusion of the unqualified from consultation, and warned against any association with quackery. However there was no hard and fast boundary; the orthodox and the registered professions were not coterminous. Since the GMC was not allowed to enforce any particular theory of medicine or system of therapeutics, the position of registered practitioners who were homeopaths (for example) was particularly troubling, especially for the BMA during the mid-nineteenth century. Opinion was divided as to whether to take a liberal tolerant stance with relation to homeopathy, or to dismiss it as dangerous heresy (Horner, 1995). The BMA's guidance on consultation stressed the 'uselessness' of calling in a homeopath, and the net effect of their advice was to leave homeopaths in a kind of tolerated limbo of professional isolation (Minty, 1932). During the early twentieth century the GMC's pronouncements on 'covering' became extended, largely through one prominent case, to effectively debar doctors from collaborating with any independent unregistered practitioners. The famous case of Dr Axham – who was erased from the *Register* in 1911 for anaesthetising patients for the famous bonesetter Herbert Barker – brought this ideal of non-collaboration in for harsh public criticism. Barker was knighted in 1924, and George Bernard Shaw took it on himself to campaign for Axham's restoration whilst simultaneously attacking the GMC and BMA as self-interested and anti-social (Morrice, 1994). This demonstrated, amongst other things, the popularity of unregistered and unorthodox practitioners with the public. This fact baffled many medical writers on the subject, who tended to see it as a sign of public ignorance, but during the 1920s a more nuanced approach to 'other practitioners' became apparent.[39]

As Bynum and Porter (1987) have pointed out, 'quackery' is in many ways a by-product of the self-proclaimed 'orthodoxy' of another group of practitioners. But the processes of professionalisation could be, and were, imitated by other practitioners in the medical marketplace, and the relationship between the medical profession and the huge variety of independent and semi-independent practitioners provided continuing ethical trouble. By the early twentieth century there appeared a number of distinct issues within this general theme. The first was the need for the attending practitioner to remain in control of the work of auxiliaries such as nurses and physiotherapists. There was also the question of to what extent other practitioners might appear to be traders or quacks. Interestingly the ethical codes of many professionalising groups such as physiotherapists (Barclay, 1994) and chiropodists actually banned behaviours disapproved by the medical profession, whilst others, such as opticians, attempted to professionalise in the teeth of medical opposition (Mitchell, 1982). One of the key issues within these other ethical codes was that of advertising, which in truth, few of these emerging professionals could afford to do without.

Advertising

In advertising we find perhaps the most protean, but fundamental, of all professional ethical issues, and one that came closest to defining what doctors were and were not. But advertising, by providing information (and disinformation) about goods and services, and seeking the greatest market share for them, is of course integral to the market economy. Through various forms, from the crying of wares to organised media campaigns, it has always sought to 'animadvert' – attract attention. Press advertising had from the beginning been associated with nostrum peddling and abortioneering, and two of the terms of abuse used by the 'orthodox' about their competitors – 'quack' and 'mountebank' – refer to attempts to attract attention. Advertising was about trade and strident claims, and thus profoundly ungentlemanly. Distaste for advertising increased during the nineteenth century, and doctors were not alone in this. The ideal espoused by doctors, along with many trades-people, was to make their way by excellence and word of mouth recommendation. The world of commerce gradually came to rely increasingly on advertising, as did those doctors struggling at the lower end of the income range (Bishop, 1949; Turner, 1952). The leaders of the profession however set themselves ever more firmly against it, in any form whatsoever. Thus as the profession became more overcrowded, financially pressured, and as contract practice organisations became more commercial in their approach to recruiting subscribers, the GMC was induced to issue a warning notice on advertising in 1899.[40] (During the high noon of government through free-market ideology, 90 years later, and encouraged no doubt by the critique of medical control of the market in healthcare, the Monopolies and Mergers Commission found the GMC's ban to operate against the public interest.[41])

The contemporary literature was at pains to suggest that there was more to the injunctions against advertising than the need to prevent unfair competition. It was presented as inimical to the nature of medical professionalism, damaging to the profession's social status, opposed to the ideal of disinterested success, and vital in distinguishing legitimate practitioners from tradesmen and quacks. Furthermore, laymen were held to be unable to evaluate the claims made about treatments. These were to be discussed within the profession, thus furthering the advancement of science and therefore the public interest.[42] By the early Edwardian era, only brass plates, with no statement as to specialisation, were allowed. Nevertheless, advertising of the individual doctor in order to attract patients, whether deliberate, indirect, inadvertent or simply perceived by sensitive colleagues, was a constant problem for the CEC. In 1905 the BMA pressed the GMC to routinely publish its Warning Notices, which included advertising and canvassing (Smith, 1993). Later, in 1923–24, the BMA persuaded the Council to extend the meaning of advertising, to include 'indirect' forms, such as newspaper articles. This latter form of advertising was mentioned by de Styrap, but became a particular problem in the 1920s, with the involvement of medical practitioners – many of them very prominent – in the popular press, often under the rubric of health education (Morrice, 1994). The BMA's supposition that such journalistic appearances could succeed in attracting patients was in fact correct.[43]

For those selling medicines, or running clinics and other commercial medical ventures, the problem was even more acute. Many in this market were not doctors, and even for medical practitioners total abstinence from advertising was out of the question. But the involvement of named doctors in these ventures, the health and cultural effects of these goods and services, and their impact on the referral relationships between practitioners made ethical debates about the involvement of doctors in such advertising inevitable. Here the BMA had a valuable asset in its ownership of the *British Medical Journal* (Bartrip, 1990). Whilst the *BMJ* depended to a great extent on advertising revenue, the threat or withdrawal of advertising copy from the *BMJ* unless the advertiser agreed to advertise only in the medical press appears, along with the disciplinary and ethical machinery, to have been an effective if crude point of influence.[44] In this second category of advertising ethics, two broad issues emerge: the questionable propriety of having an interest in something which you might recommend to your own patients, and the disruption of 'normal' relationships between doctors and patients created by commercialised medicine.[45]

Secrecy

The last great issue was that of confidentiality, or as it was then called, professional secrecy. Concern with secrecy was almost always connected with matters of sexual impropriety, venereal disease, the horrible results of illegal abortion, illegitimate pregnancy, and proceedings for divorce. Doctors were in possession of socially explosive information about their patients – particularly those from the middle and upper classes. Prior to the 1910s doctors were encouraged to be tactful, and to avoid revealing damaging information about patients. A legal precedent of 1776 had established that doctors were required to reveal information about patients when requested by a court of law. Misinterpretation of the Kitson vs Playfair slander trial of 1897 led doctors to believe that society expected total secrecy from them (Kitchin, 1941; McLaren, 1993). This conflicted with the wish to exploit information held by doctors, particularly about abortion and venereal disease in the fight against these problems. This issue was hotly debated within the BMA during the years 1915 to 1922. A series of judicially enforced disclosures – apparently made in order to quickly resolve divorce hearings, which were rising in number – triggered what can only be described as a hysterical defence of 'Hippocratic', 'immemorial' customs.[46] These traditional ideas were thought to indicate that confidences could only be broken with a patient's consent, and the ARM passed a number of problematic resolutions stating this, and attempting to limit the further use of medical secrets by the judiciary and government. Indeed consent received more attention from the BMA in connection with secrecy than it ever did in connection with surgical or other treatment. Perhaps social ruination was regarded as a graver threat to patients, and thus to doctors, than physical harm. Those within the profession who wished to see a 'privilege' for doctors in courts of law, most notably Lord Dawson (Dawson, 1922), were in fact opposed by the CEC led by Langdon-Down. Although their objections were swept aside by the Council and ARM, the BMA's new policy quickly ran into the unmitigated opposition and derision of the legal profession (Smith, 1922) and medical testimony in courts of law rose inexorably.

It is worth looking again at the ideas of 'honour and interests' here. Many doctors regarded it as offensive to their honour to reveal confidential information in court. However, the same doctors often revealed just such information in private, when the same sense of honour prompted them to do so.[47] Doctors were also happy to discuss details of patients' cases in court in order to defend the reputation and honour of other doctors accused of malpractice.[48] On one level this was a fight for influence and control between two professions, and this was widely acknowledged at the time. It might also be that the medical profession was keen to promote itself as having a monopoly on confessional medicine, a valuable claim in the medical marketplace. One commentator joked that an advertisement stating that such and such a doctor would never reveal anything would be a great boon to the doctor concerned.[49] Thus once again honour and interests were both implicated in confidentiality, in a way which a modern reader, looking at it as an issue based on rights and autonomy, would miss.

Medical Ethics before and after the Second World War

So how can we characterise early twentieth century medical ethics? It could be briefly stated to consist largely of a virtue ethic focused on gentlemanly values, particularly group and individual honour, mediated by duties to fellow professionals. It was dedicated to distancing medicine from any form of trade, and establishing gentlemanly disinterested norms of medical behaviour. Medical ethics was seen as being based on traditions and customs, and when, as in the case of secrecy, the profession felt particularly strongly, on 'immemorial', 'Hippocratic' principles. Robert Saundby characterised medical ethics as being based on three principles. He said:

> In the relation of a medical practitioner towards his colleagues, he should obey the golden rule, … 'Whatsoever ye would that men should do to you, do ye even so to them' (*St Matthew*, vii. 12); in his relations to his patients, their interests should be his highest considerations – 'Aegroti salus suprema lex'; in his relation to the State, to the laws of his country, and his civic duties, there is no better guiding principle than the words of the Gospel, 'Render, therefore, unto Caesar the things that be Caesar's' (*St Luke*, xx. 25). (Saundby, 1907:1)

Here it is important to stress the second component of this triad – acting in the patient's best interests. This was seldom discussed, but when it was, it was described as absolutely fundamental, a core definition of the doctor. It appears to have been assumed that what passed between the individual doctor and patient was governed not only by this most basic and unshakable paternalistic assumption, but also by a shared moral framework. In this feature of early twentieth century medical ethics we find not only a reason for the apparent absence of the themes that preoccupy ethicists today, but perhaps one of the reasons why these earlier professional ethics were superseded. Doctors increasingly came to be seen to have failed to pay enough attention to what 'the patients' interests' might be, and the moral consensus of the earlier period was increasingly seen to have broken down.

It is worth noting that all the principles cited by Saundby were expressed in forms that stressed antiquity, and that the Hippocratic Oath was missing from his treatment. In being most commonly referred to as the 'customs' or 'traditions' of the profession, medical ethics bore an interesting relationship with English Law. The English legal tradition is grounded in tradition and precedent, as well as legal principles and specific codification, and the British Constitution is largely unwritten. Thus an essentially unwritten, though not undescribed medical ethics based on customs and traditions represented a guide to medical behaviour that was as authoritative and adaptable as the British legal and constitutional system. This was also reflected in the casuistic approach of the CEC and the GMC, and the resistance to written codes and the consideration of hypothetical issues or scenarios. Echoing the clinical method of the time, which stressed the importance of context and environment, cases were decided with reference to all the particular circumstances, and past local practices.

Gentility and the pretence of having no interest in profit were also key elements in professional self-definition and jealously guarded by the ethical rules and conventions of the day. This came out in the rules on advertising, with the profession choosing to become conspicuous by their inconspicuousness in the medical marketplace. In other words, they sought to draw attention to the fact that they did not draw attention to themselves, and to profit by appearing not to seek profit. Patients were supposed to find their way to the doctor of their choice by force of the doctor's skill and virtue alone, and all parties were to pay scrupulous respect to this doctor-patient dyad. Many members of the public did not understand or believe in these subtleties. The rules on consultation and non-co-operation with unqualified practitioners were attacked as self-interested, and doctors were often seen as excessively concerned with income.

The ideal of medical behaviour was often referred to as 'playing the game'. British doctors were thought to have a natural tendency to this sportsmanship and sense of 'fair play'.[50] 'The Game' means a ritualised and gentlemanly contest in which the rules are agreed and upheld not only in letter but most importantly in spirit. In many ways the medical conduct structures built up by British doctors can be seen to ensure that medicine resembled a kind of massive game of cricket, or athletic contest. In this fantasy of a sunlit Corinthian idyll, the entire event was carefully modulated to allow 'the best man' or 'side' to win by merit of skill and virtue, and to let the glory reflect on all those who took part. But this arena was predicated on the exclusion of others – quacks, cheats, the unscientific, the money grubbing, and those who had 'let the side down'. Ideally the rules were to be internalised, and based on more important personal virtues. Medical ethics functioned not only to strengthen and patrol the boundary of the profession (stating who could and could not join in 'the game') but also regulated the space within it (the way the game was played).

Thus we see how medical ethics, as French has it, 'characterised', and in fact, defined 'the group'. How well did it 'contribute to its success'? Evidently, many doctors held these rules and principles to be important enough to justify short term or specific episodes of inconvenience or difficulty. For many doctors struggling to make ends meet they may have been an intolerable burden, and there are a few notable occasions in which these principles brought the BMA and the profession into fairly stark disrepute. This, along with the opprobrium with which the transgressive were treated, serves to point up the moral sentiment underlying professional 'etiquette'. On the other hand this set of ideas and rules certainly did a great deal to bolster the position of ordinary family doctors in Britain, along with many of the other countries of the former British Empire in which the BMA had branches. However, by the 1940s it was becoming clear that these traditional ideas were no longer, alone, enough to safeguard trust in doctors and the social position enjoyed by the profession. Thus the re-working of the Hippocratic Oath, and the specific codifications around human experimentation were invoked and promoted. As medical attitudes to reproduction and terminal illness began to move away from the traditional Christian attitudes, and new techniques gained ground, those seeking to defend these values, most notably members of the Roman Catholic Church contributed to an outpouring of writings about medical ethics during the 1950s and 1960s.[51]

Traditional medical ethics lived on, most notably in the work of the CEC. However, the secure working environment of the National Health Service removed many of the marketplace insecurities that had pre-occupied doctors prior to 1948, particularly General Practitioners. These factors almost certainly led to the decreasing importance and status of the CEC, and to the concerted efforts to reform its work in the late 1980s.[52] This process saw the introduction of lay members, and the new territory was marked out by a work of perhaps more political and ideological significance than relevance to the BMA's membership: a report on medical involvement in torture (BMA, 1992). The title of this report, *Medicine Betrayed*, rather than 'patients' or 'humanity betrayed', is significant. It points up the harm done to the profession by abuses of human rights and the need to rectify this in the deployment of ethical discourse.

It is in this focus on human rights, and their abuse, that we see something characteristic about modern medical ethics, in which issues of autonomy feature strongly. It would be easy to conclude that if the materials discussed in this chapter tell us anything about earlier professional ethics and autonomy, it is that it was the doctors' autonomy they most obviously sought to protect. This is most apparent in the profound and assumed paternalism of the discourse about patients' treatment. Even the issue of confidentiality, when viewed in historical context, was more to do with issues of professional honour and the need to make special claims and avoid social disgrace, than about the rights of patients. The issue of abortion was not viewed as an area in which the patient might be allowed to decide, but one in which doctors should, in consultation with one another, be allowed to decide. However, the power of patients in the pre-war medical marketplace, simply through being the doctor's customers in a partially regulated marketplace, should not be left out of this assessment. This applied in contract practice arrangements, in National Insurance (which incorporated the notion of 'free choice of doctor' at the BMA's insistence) as well as in private practice. Indeed the dependence of doctors on the autonomous choices made by patients in the medical marketplace was a major driving force in medical politics and ethics. The effort doctors put into trying to control this marketplace, their own behaviour, and through this the behaviour of patients, was in itself a sign of how threatened they had felt themselves to be. By mid-century however, the orthodox medical profession had more or less captured the market in healthcare. The timing of this achievement to coincide with all the technological and social changes of the last 50 years exposed the profession to a new threat – the harmful effects of its own hegemony, power, and culture of paternalism.

Conclusions

Medical ethics can thus be seen as an integral part of medicine's strategy of professionalisation, by adjudicating between right and wrong medical behaviour where this is not defined by law. It plays a key part in the continual re-negotiation of the social contract between profession, patients and society, in the definition of the characteristics and role of doctors, and in the definition of who and what lies beyond the medical pale. Whilst the way in which attempts to maintain and promote the position of medicine have shifted as the medical and social context shifted, this basic formulation arguably holds true in all periods. The modern situation, in which an emerging group of career ethicists has a key role in defining the content of medical ethics, reflects the modern trend to reduce the autonomy of all professions.[53] (This analysis must also relate to the nascent profession of bioethics itself, which will also have to establish the worth of its specialist skills and negotiate a role in society, a process in which voices have already been raised in criticism and scepticism.[54]) The development of professionalising bioethics also reflects the moral dilemmas created by modern medical and biotechnologies within a society that is fundamentally secular, multi-cultural, and profoundly industrialised. This is a society that at once generates and disputes new moral decisions, many of them at the interfaces of individuality, identity, death, birth, conception, and terminal illness – interfaces with which medicine so often deals. British doctors of the early twentieth century not only had limited powers in these situations, but were also working within a society in which basic moral frameworks relating to these aspects of life were more widely agreed. These modern issues were therefore not felt to affect the profession as deeply as the questions dealt with in this chapter. But in both kinds of medical ethics doctors' behaviour, moral principles, collective and individual moral choice, and the role of the profession in society were all implicated.

Notes

References to BMA internal documents are given in relation to the meeting at which they were considered. Where not otherwise stated references are to the Minutes. All these documents can be located at the British Medical Association Archive (BMAA) using their clear cataloguing system. Where references are made to ephemera, dossiers on particular issues, or other material only available at the Wellcome Trust Library's Contemporary Medical Archive Centre (CMAC), references include the appropriate CMAC catalogue number.

Council = BMA Council
CEC = Central Ethical Committee
s/c = sub-committee

1 Carr-Saunders and Wilson, 1933, pp. 90f.
2 See Mitscherlich and Mielke, 1949, pp. xxiii–xxv.
3 Declaration adopted by the General Assembly of the World Medical Association at Geneva, September 1948.
4 The first such lay involvement came with the appointment of Sir Edward Hilton Young, later Lord Kennett, to the General Medical Council in May 1926, *The Times*, 9 July 1926, p. 10b. All Privy Council appointees up to that point had been doctors.
5 See Saundby, 1907, p. 1.
6 For instance the editor of the *BMJ* described most pre-war ethical concerns as 'etiquette' in Clegg, 1957, pp. 31–45. See also Freidson, 1975, p. 245.
7 Cooter, Roger, 1998, seminar, London School of Tropical Medicine and Hygiene (and personal communication). See also Cooter, 2000, pp. 451–68.
8 Tallis, Raymond, 'The Reluctance to Trust in Trust', *Times Literary Supplement*, 30 January 1998, pp. 5f.
9 Percival, 1803, p. 69.
10 See Forbes, 1955. Forbes had served as both secretary and member of the CEC, and became Secretary of the MDU and member of the GMC. He felt Percival's codification had set out the fundamental features of medical ethics. See also Leake, 1927, which was published largely to denounce the ethical codes of the American Medical Association.
11 An example of this is to be found in Memorandum, CEC s/c, 19 January 1937.
12 RCP Bye-law XXIII.x, in RCP, 1862–1939.
13 *Trans. PMSA*, 1833, vol. 1, pp. 24f.
14 De Styrap, Jukes, 1870, 1874, 1888, 1890, *The Medico-Chirurgical Tariffs Prepared for the [late] Shropshire Ethical Branch of the B.M.A.*, private imprint, Shrewsbury, H.K. Lewis, London.
15 De Styrap, 1890, 3rd edn, p. 54.
16 Ibid., pp. 49f.
17 Bartrip, 1996, p. 143.
18 Sprigge, 1905, pp. 53ff.
19 Carr-Saunders and Wilson, 1933, p. 91.
20 Little, 1932, p. 84.

21 BMA Council, July 1903, BMAA B/54/2/11.
22 CEC s/c, 4 May 1906.
23 These statements are drawn from an analysis of attendances at all CEC and subcommittee meetings 1902-1939, and information gleaned from the *Medical Directory* and *BMJ* obituaries. See Morrice, 1999, pp. 45–59.
24 Memorandum CEC s/c, 20 February 1924.
25 A very brief note on novel treatments appears in Saundby's *Medical Ethics*.
26 Towards the end of the 1930s advice about consent to sterilising operations became apparent in the advice literature of the defence organisations and medical legal texts, but it appears to have been construed as an issue of medical defence, rather than ethics.
27 CEC, 8 December 1903.
28 For example, Memorandum, CEC, 27 May 1931.
29 For example, ARM 1933, minute 22.
30 Council, 1 October 1903.
31 *BMJ*, 1905, vol. 2, *supplement*, pp. 95f.
32 See 'Rules Governing Procedure in Ethical Matters of a *Division* not itself a Branch. Approved by the Representative Body, July 1912', *BMJ*, 1912, vol. 2, *supplement*, pp. 231–3.
33 'Joint Opinion on "Warning Notice"', 28 January 1914 in CMAC, SA BMA D183.
34 *BMJ*, 1918, vol. 2, *supplement*, 26 October 1918, pp. 53–60.
35 Patients were often referred to as female, particularly by Saundby.
36 Little, 1932, p. 292.
37 Their report was never ratified in all details by an ARM, but was issued as a 'Report on Ethics of Medical Consultation (as amended in accordance with the instructions of the Annual Representative Meeting 1909)'. See CMAC SA BMA D248.
38 Saundby implacably opposed his former colleagues on this question, see letter – printed under: 'The Ethical Aspects of Medical Consultation', *BMJ*, 1907, vol. 1, p. 534.
39 See Dawson, Bertrand (Lord Dawson of Penn), 'Those other practitioners', *BMJ*, 1928, vol. 1, p. 321.
40 GMC, *Minutes*, 1899, vol. 26, p. 275.
41 Horner, 1995, p. 81.
42 Saundby, 1907, p. 7; Sprigge (ed.), 1928, p. 8.
43 See 'Cancer clinics, the Bendien treatment, Correspondence 1933–1935', CMAC SA BMA D130.
44 See 'Advertising in connection with nursing homes, electrotherapeutic and other institutions of a similar nature. 1920–50', CMAC SA BMA D234.
45 See 'Memorandum on proposed therapeutic institutes', CEC 13 November 1928.
46 See the account of the ARM of 1921 in *BMJ*, 1921, vol. 2, *supplement*, 23 July 1921, p. 38.
47 See the account of the ARM of 1920 in *BMJ*, 1920, vol. 2, *supplement*, p. 10.
48 See letter, Hempson to Cox, 3 April 1922, in CMAC SA BMA D170.
49 Riddell, George (Lord Riddell), 'Should a Doctor Tell?', *John O'London's Weekly*, vol. 17, 16 July 1927, pp. 441–3.
50 For an example see anon., 1928, 'Medical Ethics', *BMJ*, 1928, vol. 1, p. 984, a review of Leake, 1927.
51 For two good examples see: Marshall, John, 1960, *The Ethics of Medical Practice*, Darton Longman and Todd, London; and Flood, Dom. Peter, 1953 and 1956, *New Problems in Medical Ethics*, Mercier, Cork.
52 See Horner, 1995.
53 Abelson, Julia, Maxwell, P.H. and Maxwell, R.J., 'Do Professions Have a Future?', *BMJ*, 1997, vol. 315, p. 382.

54 Anon, 'The Ethics Industry', *Lancet*, 1997, vol. 350, p. 897, styled much modern medical ethics as 'quaint and irrelevant'.

References

Archives

British Medical Association Archive (BMAA), Tavistock Square, London WC1H 9JP.
Contemporary Medical Archive Centre (CMAC), Wellcome Trust Library, 183 Euston Road, London NW1 2BN.

Published Works

Baker, R. (1993), 'Deciphering Percival's Code', in idem, Porter, D. and Porter, R. (eds), *The Codification of Medical Morality*, vol. 1: *Medical Ethics and Etiquette in the Eighteenth Century*, Kluwer Academic Publishers, Dordrecht, pp. 179–211.

Baker, R. (ed.) (1995), *The Codification of Medical Morality*, vol. 2: *Anglo-American Medical Ethics and Medical Jurisprudence in the Nineteenth Century*, Kluwer Academic Publishers, Dordrecht.

Barclay, J. (1994), *In Good Hands: the History of the Chartered Society of Physiotherapy 1894–1994*, Butterworth Heinemann, London.

Bartrip, P. (1990), *Mirror of Medicine. A History of the BMJ*, Clarendon Press, Oxford.

Bartrip, P. (1996), *Themselves Writ Large: the British Medical Association, 1832–1966*, BMJ, London.

Berlant, J. (1975), *Profession and Monopoly. A Study of Medicine in the United States and Great Britain*, University of California Press, Berkeley.

Bishop, F.P. (1949), *The Ethics of Advertising*, Robert Hale, London.

Brackenbury, H. (1935), *Patient and Doctor*, Hodder and Stoughton, London.

BMA (1949), *Ethics and Members of the Medical Profession*, BMA, London.

BMA Working Party (1992), *Medicine Betrayed: The Participation of Doctors in Human Rights Abuses*, Zed Books and BMA, London.

Bynum, W.F. and Porter, R. (eds) (1987), *Medical Fringe and Medical Orthodoxy 1750–1850*, Croom Helm, London.

Carr-Saunders, A.M. and Wilson, P.A. (1933), *The Professions*, Clarendon Press, Oxford.

Clark, Sir George (1964, 1966), *A History of the Royal College of Physicians of London*, vols. 1–2, Clarendon Press, Oxford.

Clegg, H. (1957), 'Professional Ethics', in Davidson, M. (ed.), *Medical Ethics, a Guide to Students and Practitioners*, Lloyd-Luke Ltd, London, pp. 31–45.

Cooke, A.M. (1972), *A History of the Royal College of Physicians of London*, vol. 3, Clarendon Press, Oxford.

Cooter, R. (1995), 'The Resistible Rise of Medical Ethics', *Social History of Medicine*, vol. 8, pp. 257–70.

Cooter, R. (2000), 'The Ethical Body', in idem and Pickstone, J. (eds), *Medicine in the Twentieth Century*, Harwood Academic Publishers, Amsterdam, pp. 451–68.

Cox, A. (c. 1950, not dated), *Among the Doctors*, Christopher Johnson, London.

Dawson, B., Lord Dawson of Penn (1922), 'An Address on Professional Secrecy', *Lancet*, 1922, vol. 1, p. 619.

De Styrap, J. (1878, 1886, 1890, 1895), *A Code of Medical Ethics: with Remarks on the Duties of Practitioners to their Patients, etc.*, J. and A. Churchill, H.K. Lewis, London.

Doyal, L. and Gillon, R. (1998), 'Medical Ethics and Law as a Core Subject in Medical Education', *British Medical Journal*, vol. 316, pp. 1623f.

Forbes, R. (1955), 'A Historical Survey of Medical Ethics', *St. Bartholemew's Hospital Journal*, vol. 59, pp. 282–6, 316–9.

Freidson, E. (1975), *Doctoring Together: A Study of Professional Social Control*, Elsevier, New York.

French, R. (1993), 'The Medical Ethics of Gabriele de Zerbi', in Wear, A., Geyer-Kordesch, J. and French, R. (eds), *Doctors and Ethics: the Earlier Historical Setting of Professional Ethics*, Rodopi, Amsterdam, pp. 71–94.

General Medical Council (1859ff.), *Minutes*, GMC, London.

Gisbourne, T. (1794), *An Enquiry into the Duties of Men in the Higher and Middle Classes of Society in Great Britain, Resulting from their Respective Stations, Professions and Employments*, 2 vols, B. and. J. White, London.

Green, D.G. (1985), *Working Class Patients and the Medical Establishment: Self Help in Britain from the Mid-Nineteenth Century to 1948*, Gower, Aldershot.

Harper, C.J.S. (1912), *Legal Decisions under the Medical and Dentists Acts*, Constable and Co., London.

Harris, J. (1993), *Private Lives, Public Spirit: a Social History of Britain, 1870–1914*, Oxford University Press, Oxford.

Hawthorne, C.O. (1935), 'General Practice No. IV – Medical Ethics', *Practitioner*, vol. 137, pp. 646–56.

Horner, J.S. (1995),'Medical Ethics and the Regulation of Medical Practice, with Particular Reference to the Development of Medical Ethics within the British Medical Association 1832–1993', MD thesis, Victoria University of Manchester.

Jonsen, A. (1990), *The New Medicine and the Old Ethics*, Harvard University Press, Cambridge.

Jonsen, A. (1998), *The Birth of Bioethics*, Oxford University Press, Oxford.

Kitchin, D.H. (1941), *Law for the Medical Practitioner*, Eyre and Spottiswode, London.

Lawrence, C. (1985), 'Incommunicable Knowledge: Science, Technology and the Clinical Art in Britain, 1850–1914', *Journal of Contemporary History*, vol. 20, pp. 503–20.

Leake, C.D. (1927), *Percival's Medical Ethics*, William and Wilkins Co., Baltimore.

Little, E.M. (c. 1932, not dated), *History of the British Medical Association 1832–1932*, BMA, London.

McGregor, O.R. (1957), *Divorce in England: A Centenary History*, Heinemann, London.

McLaren, A. (1993), 'Privileged Communications: Medical Confidentiality in Late Victorian Britain', *Medical History*, vol. 37, pp. 129–47.

Minty, L. le Marchant (1932), *The Legal and Ethical Aspects of Medical Quackery*, William Heinemann, London.

Mitchell, M. (1982), *History of the British Optical Association, 1895–1978*, The Association and the British Optical Association Foundation, London.

Mitscherlich, A. and Mielke, F. (1949), *Doctors of Infamy: the Story of the Nazi Medical Crimes*, Henry Schuman, New York.

Morrice, A.A.G. (1994), '"The Medical Pundits": Doctors and Indirect Advertising in the Lay Press, 1922–1927', *Medical History*, vol. 38, pp. 255–80.

Morrice, A.A.G. (1999), '"Honour and Interests": Medical Ethics in Britain, and the Work of the British Medical Association's Central Ethical Committee, 1902–1939', MD thesis, University of London.

Nye, R.A. (1995), 'Honor Codes and Medical Ethics in Modern France', *Bulletin of the History of Medicine*, vol. 69, pp. 91–111.

Percival, T. (1803), *Medical Ethics; or a Code of Institutes and Precepts Adapted to the Professional Conduct of Physicians and Surgeons*, J. Johnson and R. Bickerstaff, Manchester.

Perkin, H. (1989), *The Rise of Professional Society: England since 1880*, Routledge, London and New York.

Peterson, M.J. (1984), 'Gentlemen and Medical Men: the Problem of Professional Recruitment', *Bulletin of the History of Medicine*, vol. 58, pp. 457–73.

Pickstone, J.V. (1993), 'Thomas Percival and the Production of Medical Ethics', in Baker, R., Porter, D. and Porter, R. (eds), *The Codification of Medical Morality*, vol. 1: *Medical Ethics and Etiquette in the Eighteenth Century*, Kluwer Academic Publishers, Dordrecht, pp. 161–78.

Roberts, H. (1937), *Medical Modes and Morals*, Michael Joseph Ltd, London.

Robertson, W.G.A. (1921), *Medical Conduct and Practice, a Guide to the Ethics of Medicine*, A. and C. Black, London.

Royal College of Physicians of London (1862, 1886, 1933, 1939), *The Charter, Bye-Laws and Regulations of the Royal College of Physicians*, London.

Saundby, R. (1902), *Medical Ethics, a Guide to Professional Conduct*, John Wright, Bristol.

Saundby, R. (1907), *Medical Ethics, a Guide to Professional Conduct*, Charles Griffin, London.

Smith, Frederick Edwin, Earl of Birkenhead (1922), *Points of View*, vol. 1, Hodder and Stoughton, London.

Smith, Russell G. (1993), 'The Development of Ethical Guidance for Medical Practitioners by the General Medical Council', *Medical History*, vol. 37, pp. 57–67.

Smith, Russell G. (1994), *Medical Discipline: The Professional Conduct Jurisdiction of the General Medical Council, 1858–1990*, Clarendon Press, London.

Sprigge, S. Squire (1905), *Medicine and the Public*, Heinemann, London.

[Sprigge, Sir Samuel Squire], 'The editor of the Lancet and expert collaborators' (1928), *The Conduct of Medical Practice*, Lancet, London.

Thompson, F.M.L. (1988), *The Rise of Respectable Society: A Social History of Victorian Britain 1830–1900*, Fontana, London.

Toulmin, S. (1982), 'How Medicine Saved the Life of Ethics', *Perpectives in Biology and Medicine*, vol. 25, pp. 735–50.

Turner, E.S. (1952), *The Shocking History of Advertising*, Michael Joseph, London.

Waddington, I. (1975), 'The Development of Medical Ethics – a Sociological Analysis', *Medical History*, vol. 19, pp. 36–51.

Warnock, M. (1966), *Ethics since 1900*, 2nd edn, Oxford University Press, London.

Chapter 2

The Emergence of Medical Professional Ethics in Germany

Andreas-Holger Maehle*

Introduction

If we could ask a German doctor of the late nineteenth century 'What is medical ethics?', he would probably answer with a list of rules and duties. He would say that a doctor must not advertise his medical services; that he has to be honest in using specialist titles; that it is wrong to attract patients by offering free treatments or by undercutting other doctors' contracts with health insurers; that one must not prescribe patent medicines of unknown composition, the so-called 'secret remedies'; that it is dishonourable to poach a colleague's patients; and that one must never make negative remarks about a colleague in public. If asked why these rules had to be observed, he would probably say that 'collegiality' demanded this and that compliance was a matter of 'professional honour'. If we asked him what happened if he disregarded these rules, he would tell us that his colleagues might well bring him in front of a tribunal of his local medical society or of the chamber of physicians of his region. And that this would damage his reputation and his practice.

In fact the rules which have here been put in the mouth of an imaginary nineteenth-century doctor were the 'Principles of a Medical Professional Code', which were adopted in 1889 by the annual assembly of the German Medical Association (*Deutscher Ärztevereinsbund*) in Braunschweig (Ärztetag, 1889). At that time the German Reich had 352 medical societies, of which 135 had a professional code and 194 a disciplinary tribunal (Graf, 1890). Moreover, by the turn of the century, several German states had introduced official medical courts of honour. Attached to the state-authorised chambers of physicians or to medical district societies they were composed of doctors, lawyers, and civil servants, and judged cases of alleged professional misconduct. Braunschweig started with such courts in 1865, followed by Baden in 1883, Hamburg and Bavaria, both in 1895, the Kingdom of Saxony in

* I gratefully acknowledge the support given my research by the Wellcome Trust, the British Council, and the University of Durham (Research Committee).

1896, and Prussia in 1899. In the period leading up to the First Word War, the Prussian medical courts of honour dealt with about three accusations of misconduct per every 100 practitioners each year, most of them made by other doctors (Maehle, 1999). About one third of the cases resulted in 'punishment' which could consist of an official warning or reprimand, a fine, or withdrawal of the right to vote for, and to be elected to, the chamber of physicians. In serious cases the courts could decide to 'increase' punishment through publication of details in a medical journal. They were not entitled, however, to withdraw a doctor's licence to practice. Only the state authorities could do this, typically in cases of criminal conviction linked with the loss of civil rights (Altmann, 1900; Kade, 1906; Maehle, 1999; Übelhack, 2002).

Obviously this scenario raises questions. Why was the formal codification of professional conduct deemed necessary, and why were institutions set up to monitor and discipline doctors? What was the rationale behind the specific forms of correct behaviour that the German Medical Association demanded? And what can be said about ethical issues in relation to patients rather than to other practitioners – issues that were not mentioned in the Braunschweig Principles, such as medical confidentiality and patients' consent to treatment?

Professional Conduct

Codes of conduct are, of course, a typical feature of professionalisation, but there were also more specific reasons for their introduction into German medicine. As in other European countries and North America, the literary tradition of medical deontology, outlining the physician's duties to colleagues, patients, and society at large, continued throughout the nineteenth century and beyond (Brand, 1977). Such literature aimed at raising the moral status of doctors in comparison with their non-licensed competitors on the medical marketplace, the so-called *Kurpfuscher*. One of the key authors of this genre, the Prussian Royal Physician Christoph Wilhelm Hufeland (1762–1836), for example, applied Immanuel Kant's categorical imperative in demanding that patients must always be seen as an end in themselves, not as a means to improve the art of medicine or to perfect scientific experimentation (Hufeland, 1836: 893). He also required that the doctor be an uncompromising protector of human life, a position that ruled out abortion and active euthanasia (Hufeland, 1836: 898). Nineteenth century doctors generally followed this strict line, especially as both abortion and 'killing on demand' were included as crimes in the German Penal Code of 1871 (Seidler, 1993; Benzenhöfer, 1999). An abortion performed by a doctor remained unpunished, however, if the pregnancy had acutely endangered the woman's life, e.g. if a natural birth had been impossible due to a narrow pelvis. Medical deontological literature also contained much practical advice for young physicians in the tradition of the eighteenth century *savoir faire*: how to gain the trust of patients, how to entertain good relations with colleagues, and how to build up a lucrative practice (French, 1993; Geyer-Kordesch, 1993; Ritzmann, 1999). However, widespread as it was, medical deontology often lacked mechanisms of enforcement in practice. As Hufeland admitted, with regard to proper treatment of patients one really had to rely on the individual conscience of doctors, their 'inner tribunal' (Hufeland, 1836: 894).

A few obligations were imposed by the state however. Prussia required through its Penal Code of 1851 that doctors provide professional assistance for any patient who requested it – whether he could pay or not, and at any time, day or night. A Doctor's Oath demanded – besides allegiance to the Prussian King – to practise conscientiously and to the best of one's knowledge. However, both the oath and the duty of medical assistance were abolished with new Trading Regulations, which had considerable influence on the development of the German medical profession (Huerkamp, 1985).

Introduced first in 1869 by the states of the North German Confederation and extended in 1871 to the whole of the newly founded German Reich, these Regulations stated that medical practice was a free trade. It could be exercised by anyone, only the use of the title *Arzt* (i.e. physician) was legally protected. While this change had actually been promoted by liberal parts of the medical profession, in particular the Berlin Medical Society, it led in the following years to a backlash of activities by more conservative doctors who saw the need to demarcate themselves sharply from non-licensed healers. Exclusive organisation in professional societies and chambers of physicians, adoption of strict codes of conduct, and discipline of doctors' behaviour by medical tribunals were all used to achieve this aim (Graf, 1890; Berger, 1896; Marx, 1907; Gabriel, 1919). Such measures became even more important as the numbers of doctors, and thus competition within the profession, grew. In the German Reich the ratio of doctors to inhabitants increased from about 1:3,200 in the mid-1880s to 1:2,200 at the turn to the twentieth century (Herold-Schmidt, 1997). Another important factor was the introduction of compulsory sickness insurance for workers in 1883.[1] On the one hand it brought patients from the lower end of the social scale into doctors' practices, but on the other hand it led to fierce competition among practitioners for lucrative contracts with health insurers (Labisch, 1997).

Interestingly, the first German medical societies that adopted a binding professional code imported it from abroad. The Munich Medical District Society decided in 1875 to make a German translation of the *Code of Ethics* of the American Medical Association (1847) their 'binding norm' (Ärztlicher Bezirksverein München, 1875; Baker, 1995). The Karlsruhe Medical Society adopted in the following year a very similar, though shorter version (Ärztlicher Kreisverein Karlsruhe, 1876). The Munich and Karlsruhe codes, in turn, both formed the basis of the 'Braunschweig Principles' mentioned above.

Medical ethics was used to fight the problem of competition, both within the profession and in relation to other healers. All of the Braunschweig Principles directly or indirectly served this aim. The prohibition of advertising is a particularly clear case in point (Binder, 2000). Blatant advertising of medical services and cures, especially in the daily press, was a typical feature of non-licensed healers, the *Kurpfuscher* or quacks, who were important competitors on the German health care market throughout the Second Reich and the Weimar Republic (Spree, 1989). The rule against medical advertising, which was issued routinely by medical professional bodies, aimed at elevating doctors above the mercenary behaviour of those other healers. Yet simultaneously this rule tried to tackle the problem of competition among qualified medical practitioners themselves. The creation of a uniform offer of medical services towards the clients, i.e. patients and health insurers, was a central objective of professional politics. This essentially economic aim was part of the striving for a monopoly of health care provision, though it was couched in the language of ethics. Advertising as well as underbidding, and the prescription of patent medicines (Woycke, 1992), were said to damage the dignity and honour of the medical profession and to violate the duty of collegiality (Ehrengerichtshof, 1908–34). The honour code of a doctor was supposed to be that of the academic and gentleman, not that of the businessman or trader. In fact the medical tribunals had been formed in imitation of the courts of honour for military officers and for lawyers. Professional honour (*Standesehre*) was central to medical ethics (Maehle, 1999; Übelhack, 2002). As the sociologist Georg Simmel (1858–1918) described it, honour formed the important middle ground between law and morality. It guaranteed proper conduct in areas that could not be reached by the law and in which morality, based on the individual's conscience, was not reliable enough (Simmel 1958: 403).

Yet many doctors apparently did not agree with this lofty ideal, or were forced to ignore it by economic necessity. Offences against the prohibition of medical advertising formed a major or even the largest part of cases of medical misconduct. The Medical Court of Honour for Brandenburg and Berlin, for example, named 'advertising' most frequently (72 times) as the reason for disciplinary punishment in its annual reports between 1903 and 1920. This was followed by financial misconduct (30 times), slander or libel (29 times), sexual offences (23 times), lack of collegiality (21 times), and business links with non-licensed healers (18 times).[2] Apart from the prosecution of sexual misconduct, all of the major issues dealt with by the courts of honour concerned the profession itself rather than the doctor-patient relationship. Also political statements by doctors could lead to punishment, if they were deemed to be offensive or dishonourable, and therefore harmful to the reputation of the profession (Maehle, 1999). One may be tempted to assume that German medical ethics around 1900 was chiefly professional etiquette – not 'real' ethics in a modern sense. However, such a view would be superficial.

For one, also apparent etiquette rules, such as those against advertising or against secret remedies, had an ethical dimension with regard to patients and the public more generally. It was at least claimed by the medical profession that such rules protected patients against financial exploitation and dangerous treatments (Binder, 2000). In a wider sense there was the claim that only academically trained doctors could provide adequate medical treatment. The aim of a professional monopoly was displayed as one that would benefit the client. Moreover there were some other issues in professional ethics that directly concerned the doctor-patient relationship.

Medical Confidentiality

One such issue was the question of the limits of medical confidentiality, or professional secrecy as it was also referred to. Confidentiality was a demand already included in the Hippocratic Oath, and it was routinely mentioned in nineteenth century professional codes. More importantly, it was a legal requirement for German doctors, surgeons, midwives, apothecaries, and their helpers, in the same way as it was for lawyers. According to § 300 of the German Penal Code of 1871 these persons could be punished with a fine of up to 1,500 Marks or imprisonment up to three months, if they disclosed 'without authorisation' so-called 'private secrets' which had been entrusted to them in their capacity as office-holders or members of a profession or trade. This regulation had numerous precursors in the legislation of the various German territories since the early eighteenth century (Liebmann, 1886; Exner, 1909; Placzek, 1909).

A typical exception for doctors was the notification of dangerous infectious diseases (i.e. leprosy, cholera, typhus, yellow fever, plague, and smallpox), which was legally prescribed in 1900. Another exception was the making of statements as a witness in court, although doctors had the right to refuse giving evidence with reference to their legal duty of professional secrecy. A severe conflict between ethics and law arose, however, in cases of venereal disease, in particular syphilis. Doctors acted illegally if they gave hints to foreseeable sexual contacts of the patient in order to prevent their infection. A frequently quoted situation was that of the father of the prospective bride who asked the doctor about the health of his future son-in-law. Some doctors openly proposed a life insurance for the future husband in such cases. If the latter then refused the necessary examination of his health, the bride's family could draw their conclusions from this (Moll, 1902: 109; Placzek, 1909: 104).

The issue was brought forward, however, by a somewhat different case, which went in 1905 to the Supreme Court of the German Reich (*Reichsgericht*). Because of its significance the relevant details need to be given here. A married woman had taken her child to a doctor to have it vaccinated. On this occasion the doctor learned that the woman and her sister-in-law lived closely together in the same flat, the woman's children also sharing a bed with the sister-in-law. Since the doctor knew that the sister-in-law was infected with syphilis, he warned the married woman to keep the children away from their aunt because the latter had, as he put it, a 'contagious disease'. The woman replied: 'I can imagine what the slut has again, she doesn't come home before 3 or 4 o'clock in the morning, and all the time she's off to see the doctor.'[3] On this the doctor remarked: 'Well, if you think it may be something such as syphilis, then you had better take care.'[4] The woman afterwards told this to a neighbour who further distributed the news. A few days later the sister-in-law and her mother went to the doctor, and the mother asked what 'rumours' he was spreading about her daughter. The doctor now declared frankly that her daughter suffered from syphilis, and a heated argument followed.

The doctor had thus twice disclosed private information on a patient to third parties. Nevertheless the Supreme Court vindicated his conduct in both instances. Concerning the first instance the court argued that the doctor experienced a collision of duties. On the one hand he had a duty to protect the woman's child from infection, and if he had ignored this duty, he could have been charged with having caused physical injury through negligence. On the other hand § 300 of the Penal Code demanded professional secrecy. The court referred in this situation to the Prussian law on medical courts of honour of 1899, which demanded in its central § 3 that doctors practise conscientiously. This, argued the court, constituted a professional duty that included the obligation to warn patients of a danger of infection. Therefore, ran the conclusion, the doctor was in a position where he could disclose the secret with authorisation, and thus did not contravene the paragraph on professional secrecy. Concerning the second instance, i.e. the disclosure of the diagnosis to the patient's mother, the court pointed out that at this stage authentic information about the patient's illness was already in the public sphere. Moreover, in seeing the doctor together with her mother the patient had implicitly given consent to disclosure.[5]

While this decision was widely welcomed in the medical profession, lawyers were divided in their opinion. The matter was even brought up in both houses of the Prussian parliament. In the upper house it prompted a remarkable statement by the director of the medical department in the Ministry of Religious, Educational and Medical Affairs, Dr Förster, who was also the chairman of the Berlin appeal court (*Ehrengerichtshof*) for the medical courts of honour in Prussia. For him, the Supreme Court's judgement could be seen from two perspectives. From the point of view of public health it was understandable. But from the perspective of the doctor-patient relationship, which he regarded as a contract based on confidentiality and trust, the decision had to be condemned. Förster feared that the decision had severely shaken this contract relationship and that patients would in future avoid seeing doctors in confidential matters (Placzek, 1909: 27–8). Nevertheless Förster acquitted in 1907, as chairman of the Berlin *Ehrengerichtshof*, a doctor who had likewise breached confidentiality. This doctor had reported a promiscuous, syphilitic teacher to the inspector of schools. Like the Supreme Court two years earlier, the tribunal used the argument of a collision of duties and stated a higher duty to the welfare of the community.[6] The debate on confidentiality thus highlighted important features of contemporary medical ethics. On the one hand there was the conception of the doctor-patient relationship as a tacit legal contract. For example, only a few years earlier, in 1902, the Berlin neurologist Albert Moll (1862–1939) had published a comprehensive handbook of medical ethics, which was entirely based on this premise. Moll explicitly rejected any foundation of medical ethics on systems of moral philosophy, such as utilitarianism, or on moral theology (Moll, 1902: 7–11). On the other hand this contract relation was challenged around the turn of the century from two sides. Firstly, there was the demand by the state that doctors served not only the well-being of the individual patient, but also the welfare of the public, and that the latter was more important than the former. This demand had found its tangible expression in the mentioned law on compulsory notification of dangerous infectious diseases in 1900. The issue became even more prominent in the Weimar Republic, with a law in 1927 on the compulsory treatment of patients with venereal diseases (Sauerteig, 2000). Secondly, the contract relation was undermined by a continuing tradition of medical paternalism. As in other European countries, this paternalism was connected with the rise of hospital medicine in the nineteenth century, which put patients in a rather subordinate position towards their doctors (Elkeles, 1996a). This position was reinforced outside the hospital through the sickness insurance schemes, because it was their panel doctors (*Kassenärzte*) who decided whether and how long a patient was paid sickness benefit (Huerkamp, 1985). Paternalism was ingrained in medical deontology. As late as 1897 the Berlin physician and medical historian Julius Pagel (1859–1912) taught medical students that the doctor had to be the 'sovereign' of his patients and that it was best to regard them as 'psychologically affected, often relapsing sinners against hygiene' (Pagel, 1897: 42, 45).

The Issue of Consent

The problems with medical paternalism and disregard for patient's decision-making became apparent in two issues towards the end of the nineteenth century: consent to clinical experimentation and consent to surgery. The late 1890s saw a public scandal around the Breslau professor of venereology, Albert Neisser (1855–1916). In order to explore the possibility of an immunisation against syphilis he had injected eight, partly minor, female hospital patients with cell free blood serum from a syphilitic patient. The test persons, who had been admitted to hospital for the treatment of other diseases, had neither been informed about the nature of the injections nor had they been asked for their consent. Four of these patients later contracted syphilis. It remained unclear whether the infection was acquired 'naturally' (all four were prostitutes) or whether it was due to the experimental injections. Neisser was disciplined with a reprimand and fine by a tribunal for civil servants, because he had failed to obtain the consent of his subjects or their guardians. Newspapers reported on similar cases in other German hospitals, and the case of Neisser was brought up in the Prussian parliament. The Prussian Minister for Religious, Educational and Medical Affairs took action in 1900 by issuing a directive to the heads of hospitals. It demanded informing and gaining the consent of human subjects in scientific, non-therapeutic trials and prohibited such trials on minors or otherwise legally incompetent persons (Elkeles 1996b; Sauerteig, 2000). The majority of the medical profession, however, remained silent on the ethics of human experimentation. The requirement of 'collegiality' and general support for a scientific approach to medicine were probably the reasons for this silence. An exception was Albert Moll who pointed to the magnitude of the problem by publishing a summary of over 600 scientific, non-therapeutic human trials that he had identified in the literature. The experiments, dating from the second half of the nineteenth century, had not only been performed in Germany, but also in several other European countries and the USA. The descriptions often failed to mention whether consent had been obtained, and – as Moll emphasised – the experimental procedures were frequently dangerous or at least involved some molestation of the test persons (Moll, 1902: 505–52). Accordingly, Moll demanded written and witnessed consent to serious experimental interventions (Moll, 1902: 567–8).[7] In the following years, however, the issue of information and consent hardly surfaced in the decisions of the medical courts of honour. Between 1900 and 1914 the *Ehrengerichtshof* in Berlin published just one case each on lack of patient information and lack of consent (Ehrengerichtshof, vol. 1, 1908: 98–100; vol. 3, 1914: 124–5).

The issue of consent to therapeutic procedures, in particular surgery, was initially part of a rather theoretical legal debate about the question whether medical interventions without the patient's consent constituted physical injury or battery in the sense of § 223 of the German Penal Code. Two major opinions developed. One, represented by the Basle penologist Lassa Oppenheim (1858–1919) and the Heidelberg professor of law, Karl von Lilienthal (1853–1927), said that medical interventions were objectively battery in the sense of criminal law and that seeking the patient's consent was therefore mandatory (Oppenheim, 1892; Lilienthal, 1899). The other view, expressed by the jurists Carl Stooss (1849–1934) in Vienna and Richard Schmidt (1862–1944) in Freiburg (Breisgau), rejected this battery theory of surgical or other therapeutic procedures and regarded information and consent as a matter of medical professional ethics, but not as a legal requirement (Stooss, 1898; Schmidt, 1900). Central to the debate became a decision of the German Supreme Court in 1894. A Hamburg surgeon had performed a medically indicated operation (resection of purulent, tuberculous bones of the foot) on a seven-year-old girl against the explicit wishes of the child's father. The surgeon was eventually acquitted of battery on a technicality, but the Supreme Court issued an important ruling: operations without consent constituted assault and battery. They were punishable, stated the Court, if the doctor could not derive his right to operate 'from an existing contractual relation or the presumptive consent, the assumed brief of duly legitimised persons'.[8] In the years before the First World War the Supreme Court further endorsed this view through its decisions on several similar cases – despite protests from the medical profession and criticism from some experts in criminal law. In a way the 1894 decision thus marked the beginning of a formal requirement of consent in therapeutic situations (Maehle, 2000).[9]

Conclusions

In Germany, medical professional ethics in the nineteenth century and the early twentieth century was characterised by distinct developments. We have seen how rules of conduct were set up and enforced by medical professional bodies in order to demarcate doctors from non-licensed healers and to defuse competition among doctors themselves. This was why issues such as advertising and collegiality were so prominent, particularly in the context of the sickness insurance system. Yet medical ethics also served to protect patients. The requirements of conscientious practice and confidentiality were crucial parts of this ethic and also had legal status. Nineteenth-century doctors generally followed an ethic of benevolent paternalism.

Yet the turn to the twentieth century brought challenges to this attitude. In the areas of surgery and clinical experimentation legal notions of the self-determination of patients gained prominence. While this might have strengthened the concept of a (tacit) legal contract between doctor and patient, contemporary jurisdiction also provided arguments for sacrificing the interest of the individual patient in the interest of the community, and thus to break this contract under certain circumstances. The classical demand on doctors, *Salus aegroti suprema lex* (i.e. the welfare of the patient shall be the highest law) was confronted with the demand *Salus publica suprema lex* (i.e. the welfare of the public shall be the highest law). The latter notion was to become more prominent after the First World War. The notorious debates on the so-called 'release of the destruction of life unworthy of life' in the Weimar Republic, which then led to the Nazi 'euthanasia' programmes of the Third Reich (Burleigh, 1994), can be seen under this aspect. Moreover, ethical guidance on clinical experimentation did not find much resonance in the medical profession. Although new guidelines extending the requirement of consent to therapeutic trials were issued by the Ministry of the Interior in 1931, they failed to prevent the atrocious human experiments in the concentration camps of Nazi Germany (Kanovitch, 1998). Medical professional ethics in Germany had largely emerged in response to economic pressures, legal decisions, and health policies of the state, rather than from intense moral debate. This origin may help to explain why it did not resist the demands and temptations of Nazism.

Notes

1 On the impact of the sickness insurance system on German medicine see chapter three by Sauerteig, below.
2 Geheimes Staatsarchiv Preußischer Kulturbesitz I. HA Rep. 76 VIII B, Nr. 830.
3 Cf. *Entscheidungen des Reichsgerichts in Strafsachen*, vol. 38, 1905, p. 63.
4 Cf. ibid.
5 Cf. ibid., pp. 63-6.
6 *Entscheidungen des Preußischen Ehrengerichtshofes für Ärzte*, vol. 1, 1908, pp. 93–8.
7 On the further development of ethical guidelines on clinical research see chapter five by Tröhler, below.
8 Cf. *Entscheidungen des Reichsgerichts in Strafsachen*, vol. 25, 1894, p. 382.
9 On the further history of consent to surgery see chapter four by Prüll and Sinn, below.

References

Altmann, F. (1900), *Ärztliche Ehrengerichte und ärztliche Selbstorganisation in Preußen*, Verlag von H.W. Müller, Berlin.
Ärztetag (1889), 'Grundsätze einer ärztlichen Standesordnung', *Ärztliches Vereinsblatt für Deutschland*, vol. 18, p. 273.
Ärztlicher Bezirksverein München (1875), *Der ärztliche Stand und das Publikum: Eine Darlegung der beiderseitigen und gegenseitigen Pflichten*, Verlag von J.A. Finsterlin, Munich.
Ärztlicher Kreisverein Karlsruhe (1876), 'Standesordnung', in Altmann, F. (1900), pp. 179–83.
Baker, R. (1995), 'The Historical Context of the American Medical Association's 1847 *Code of Ethics*', in Baker, R. (ed.), *The Codification of Medical Morality*, vol. 2: *Anglo-American Medical Ethics and Medical Jurisprudence in the Nineteenth Century*, Kluwer Academic Publishers, Dordrecht, pp. 47–63.
Benzenhöfer, U. (1999), *Der gute Tod? Euthanasie und Sterbehilfe in Geschichte und Gegenwart*, Verlag C.H. Beck, Munich.
Berger, H. (1896), *Geschichte des ärztlichen Vereinswesens in Deutschland*, J. Alt, Frankfurt/M.
Binder, J. (2000), *Zwischen Standesrecht und Marktwirtschaft: Ärztliche Werbung zu Beginn des 20. Jahrhunderts im deutsch-englischen Vergleich*, P. Lang, Frankfurt/M.
Brand, U. (1977), *Ärztliche Ethik im 19. Jahrhundert. Der Wandel ethischer Inhalte im medizinischen Schrifttum*, H.F. Schulz Verlag, Freiburg/Breisgau.

Burleigh, M. (1994), *Death and Deliverance: 'Euthanasia' in Germany c. 1900–1945*, Cambridge University Press, Cambridge.

Ehrengerichtshof (1908–34), *Entscheidungen des Preußischen Ehrengerichtshofes für Ärzte*, vol. 1–5, Verlagsbuchhandlung von R. Schoetz, Berlin.

Elkeles, B. (1996a), 'Der Patient und das Krankenhaus', in Labisch, A. and Spree, R. (eds), *'Einem jeden Kranken in einem Hospitale sein eigenes Bett'. Zur Sozialgeschichte des Allgemeinen Krankenhauses in Deutschland im 19. Jahrhundert*, Campus Verlag, Frankfurt/M., pp. 357–73.

Elkeles, B. (1996b), *Der moralische Diskurs über das medizinische Menschenexperiment im 19. Jahrhundert*, G. Fischer, Stuttgart.

Exner, B. (1909), *Das Berufsgeheimnis des Arztes gemäß § 300 des Str. G. B.*, Law thesis, Universität zu Heidelberg.

French, R. (1993), 'Ethics in the Eighteenth Century: Hoffmann in Halle', in Wear, A., Geyer-Kordesch, J. and French, R. (eds), *Doctors and Ethics: The Earlier Historical Setting of Professional Ethics*, Rodopi, Amsterdam, pp. 153–80.

Gabriel, A. (1919), *Die staatliche Organisation des deutschen Ärztestandes*, Adler-Verlag, Berlin.

Geyer-Kordesch, J. (1993), 'Natural Law and Medical Ethics in the Eighteenth Century', in Baker, R., Porter, D. and Porter, R. (eds), *The Codification of Medical Morality*, vol. 1: *Medical Ethics and Etiquette in the Eighteenth Century*, Kluwer Academic Publishers, Dordrecht, pp. 123–39.

Graf, E. (1890), *Das ärztliche Vereinswesen in Deutschland und der Deutsche Ärztevereinsbund*, Verlag von F.C.W. Vogel, Leipzig.

Herold-Schmidt, H. (1997), 'Ärztliche Interessenvertretung im Kaiserreich 1871–1914', in Jütte, R. (ed.), *Geschichte der deutschen Ärzteschaft. Organisierte Berufs- und Gesundheitspolitik im 19. und 20. Jahrhundert*, Deutscher Ärzte-Verlag, Cologne, pp. 43–95.

Huerkamp, C. (1985), *Der Aufstieg der Ärzte im 19. Jahrhundert. Vom gelehrten Stand zum professionellen Experten: Das Beispiel Preußens*, Vandenhoeck and Ruprecht, Göttingen.

Hufeland, C.W. (1836), *Enchiridion medicum oder Anleitung zur medizinischen Praxis*, 2nd edn, Jonas Verlagsbuchhandlung, Berlin.

Kade, C. (1906), *Die Ehrengerichtsbarkeit der Ärzte in Preußen. Eine Bearbeitung des Ehrengerichtsgesetzes und der veröffentlichten Entscheidungen des ärztlichen Ehrengerichtshofes*, Verlag von A. Hirschwald, Berlin.

Kanovitch, B. (1998), 'The Medical Experiments in Nazi Concentration Camps', in Tröhler, U. and Reiter-Theil, S. (eds), *Ethics Codes in Medicine: Foundations and Achievements of Codification since 1947*, Ashgate, Aldershot, pp. 60–70.

Labisch, A. (1997), 'From Traditional Individualism to Collective Professionalism: State, Patient, Compulsory Health Insurance, and the Panel Doctor Question in Germany, 1883–1931', in Berg, M. and Cocks, G. (eds), *Medicine and Modernity: Public Health and Medical Care in Nineteenth- and Twentieth-Century Germany*, Cambridge University Press, Cambridge, pp. 35–54.

Liebmann, J. (1886), *Die Pflicht des Arztes zur Bewahrung anvertrauter Geheimnisse*, J. Baer and Co., Frankfurt/M.

Lilienthal, K. von (1899), 'Die pflichtmäßige ärztliche Handlung und das Strafrecht', in Juristische Fakultät der Universität Heidelberg (ed.), *Festgabe zur Feier des fünfzigsten Jahrestages der Doktor-Promotion des Geheimen Rates Professor Dr. Ernst Immanuel Bekker*, Verlag von O. Haering, Berlin.

Maehle, A.-H. (1999), 'Professional Ethics and Discipline: The Prussian Medical Courts of Honour, 1899–1920', *Medizinhistorisches Journal*, vol. 34, pp. 309–38.

Maehle, A.-H. (2000), 'Assault and Battery, or Legitimate Treatment? German Legal Debates on the Status of Medical Interventions without Consent, c. 1890–1914', *Gesnerus*, vol. 57, pp. 206–21.
Marx, C. (1907), *Die Entwickelung des ärztlichen Standes seit den ersten Dezennien des 19. Jahrhunderts*, Verlag von Struppe and Winckler, Berlin.
Moll, A. (1902), *Ärztliche Ethik. Die Pflichten des Arztes in allen Beziehungen seiner Thätigkeit*, Verlag von F. Enke, Stuttgart.
Oppenheim, L. (1892), *Das ärztliche Recht zu körperlichen Eingriffen an Kranken und Gesunden*, B. Schwabe Verlagsbuchhandlung, Basle.
Pagel, J. (1897), *Medicinische Deontologie. Ein kleiner Katechismus für angehende Praktiker*, Verlag von O. Coblentz, Berlin.
Placzek, S. (1909), *Das Berufsgeheimnis des Arztes*, 3rd edn, Verlag von G. Thieme, Leipzig.
Ritzmann, I. (1999), 'Der Verhaltenskodex des "Savoir faire" als Deckmantel ärztlicher Hilflosigkeit? Ein Beitrag zur Arzt-Patient-Beziehung im 18. Jahrhundert', *Gesnerus*, vol. 56, pp. 197–219.
Sauerteig, L. (2000), 'Ethische Richtlinien, Patientenrechte und ärztliches Verhalten bei der Arzneimittelerprobung (1892–1931)', *Medizinhistorisches Journal*, vol. 35, pp. 303–34.
Schmidt, R. (1900), *Die strafrechtliche Verantwortlichkeit des Arztes für verletzende Eingriffe. Ein Beitrag zur Lehre der Straf- und Schuldausschließungsgründe*, Verlag von G. Fischer, Jena.
Seidler, E. (1993), 'Das 19. Jahrhundert. Zur Vorgeschichte des Paragraphen 218', in Jütte, R. (ed.), *Geschichte der Abtreibung. Von der Antike bis zur Gegenwart*, Verlag C.H. Beck, Munich, pp. 120–38.
Simmel, G. (1958), *Soziologie. Untersuchung über die Formen der Vergesellschaftung*, 4th edn, Duncker and Humblot, Berlin.
Spree, R. (1989), 'Kurpfuscherei-Bekämpfung und ihre sozialen Funktionen während des 19. und zu Beginn des 20. Jahrhunderts', in Labisch, A. and Spree, R. (eds), *Medizinische Deutungsmacht im sozialen Wandel des 19. und 20. Jahrhunderts*, Psychiatrie-Verlag, Bonn, pp. 103–21.
Stooss, C. (1898), *Chirurgische Operation und ärztliche Behandlung. Eine strafrechtliche Studie*, Verlag von O. Liebmann, Berlin.
Übelhack, B. (2002), *Ärztliche Ethik – Eine Frage der Ehre? Die Prozesse und Urteile der ärztlichen Ehrengerichtshöfe in Preußen und Sachsen 1918-1933*, Peter Lang, Frankfurt/M., in press.
Woycke, J. (1992), 'Patent Medicines in Imperial Germany', *Canadian Bulletin of Medical History*, vol. 9, pp. 41–56.

Chapter 3

Health Costs and the Ethics of the German Sickness Insurance System

Lutz D.H. Sauerteig

Introduction

In medical ethics many conflicts arise from the issue of costs. How much money should society spend and indeed, what can it afford to spend on health care? Who is going to pay for medical treatment? How should health care resources be allocated? Who decides which treatment is appropriate? These are central questions on health policy discussed from the late nineteenth century onwards.

Introduced in Germany in 1883 compulsory sickness insurance not only marked the starting point of Germany's comprehensive social security system; it also initiated the first social security system in the world (Ritter, 1991; Hentschel, 1983; Ritter, 1983; Tennstedt, 1983, 1977, 1976).

The 1883 Sickness Insurance Act had a deep impact on the health care market and on society as a whole (Alber, 1992). Social policy, of which compulsory sickness insurance was a central part, aimed at reallocating resources within society from those economically active to those who were inactive, from the healthy to the sick, from the young to the old, and from the unmarried to families (Reuter, 1980: 110–15).[1] This chapter analyses some of the major consequences of compulsory sickness insurance for the health care market in Germany.

Germany's health care system is financed through statutory sickness insurance schemes and controlled by self-administrated institutions. It represents an alternative situated between – at one end of the spectrum – government-funded and state-controlled health care systems (for example that of Great Britain) and – at the other end of the spectrum – health care systems which are largely organised as a free market economy (for example in the United States). While state-financed health care systems aim at equal distribution of resources to all their citizens, this is not the case in health care systems which are based on the consumer's choice as a guiding principle. Such health care systems are often characterised by inequality of access to health care (Anderson, 1972).

How did the German health care system work in the nineteenth and twentieth centuries? Before compulsory sickness insurance was introduced in 1883, a patient consulted his or her physician, then paid the physician directly for the treatment. If the patient needed medication, he or she obtained a prescription from the physician to go to a pharmacist and buy necessary drugs which in most cases the pharmacist had to prepare himself. If an operation was necessary which the physician was unable to undertake in his consulting room or, what was more common at that time, at the patient's home, the patient was sent to hospital where he or she had to pay for the treatment as well as for a bed, food and all other services. Before 1883 the patient had to pay for any medical service he or she received. Because of this it was mainly the upper classes and the bourgeoisie who could afford to call a physician or buy remedies.

For the growing number of the poor and labouring poor, for most of the rural population and the working class in the cities, most medical services attended by physicians and any hospital treatment were too expensive. They had to rely mostly on self-help or neighbours, on local lay healers, midwives, or pharmacists. During the first half of the nineteenth century, however, physicians were obliged to treat poor patients in cases of emergency for a lower fee or even free of charge. This obligation (*Kurierzwang*) derived from the official oath sworn by physicians and subsequently, in 1851, was empowered through the Prussian Penal Code. This obligation to treat free of charge was heavily contested by physicians and was suspended in 1869 under the Trading Regulations of the North German Federation. Afterwards an obligation only arose through adherence to the codes of physicians' professional organisations. They obligated doctors to treat anyone in case of emergency regardless of the patient's own financial means. Furthermore, voluntary hospitals run by the Protestant or Catholic churches offered free treatment for the needy. However, there were not enough denominational hospitals during the first half of the nineteenth century, and most hospitals were more or less institutions for the socially marginalised, the elderly and the mentally ill, in any case, and not hospitals in the modern sense. Only the guilds and poor law institutions provided financial help for the working class and the poor during illness. In the case of poor law institutions, however, this resulted in discrimination, such as forfeiture of one's right to vote. Whereas the southern German states such as Bavaria tended to prefer a welfare system based upon the poor law, the Prussian government established a separate welfare system.

The 1794 Prussian *Allgemeines Landrecht* made it compulsory for employers to provide welfare for their servants. Friendly societies of the guilds (*Freie Hilfskassen*) or sickness insurance funds for miners and for journeymen (*Knappschaftskassen* and *Gesellenkassen*), introduced during the first half of the nineteenth century, supported their members in case of illness. Local authorities began to establish compulsory sickness insurance funds

(*Gemeindliche Zwangshilfskassen*) and some of the larger factories founded their own sickness insurance schemes for their workers. In 1875 there were 4,763 sickness insurance funds in Prussia, the majority run by factories, with nearly 780,000 members (Übelhack, 2002, ch. III.2.3.; Loetz, 1993; Huerkamp, 1985: 254–61; Stollberg, 1983; Tennstedt, 1983; Frevert, 1981; Sachße and Tennstedt, 1980).

One of the first major steps towards Germany's future statutory sickness insurance system was the 1876 Act extending Prussian legislation to the other German states by allowing local authorities to force journeymen and factory workers to join compulsory sickness insurance schemes. Local authorities could also make factory owners pay up to a maximum of 50 per cent of their workers' sickness insurance premium (Tennstedt, 1983). Despite these early measures to provide financial help in case of sickness only a marginal 5 per cent of the German population was covered by sickness insurance by 1880. The main purpose of all sickness insurance was to substitute for loss of earnings during sickness, at least to some degree.

When the compulsory sickness insurance system was introduced in 1883, the basic principles of compulsory membership and the financial contribution of employers to the insurance premium were not new. Still, the statutory sickness insurance scheme fundamentally changed the health care market in Germany. In the beginning, the majority of insured persons were blue-collar workers, but from 1911 onwards white collar-workers were also legally required to join one of the several different sickness insurance schemes (*Krankenkassen*), provided that they earned less than a specific amount of money annually. Over time this income limit continued to rise, not only to adjust to an increase in overall income, but also to include more workers (as their income rose) in the sickness insurance system. This increased the revenue of sickness insurance schemes. Physicians generally battled against any increase in the income limit as this meant their losing private patients. Sickness insurance schemes, on the other hand, often demanded raising the income limit in order to increase the number of members with higher income.

As a result the number of workers falling under the 1883 Sickness Insurance Act grew and the proportion of insured persons increased considerably. In 1885 4.3 million Germans, approximately 10 per cent of the population, belonged to one of the many different sickness insurance schemes; by the turn of the century this number had doubled, and in 1913 the sickness insurance schemes counted some 13.6 million members, approximately 25 per cent of the population. Moreover, as an increasing number of sickness insurance schemes also made their benefits available to the families of their members, it is estimated that before the outbreak of the First World War about 50 per cent of the German population had statutory sickness insurance coverage.

In 1941 the Nazis included pensioners in the statutory sickness insurance system, thus dismissing the principle connecting sickness insurance to employment. This was a crucial decision for the future of sickness insurance, because in order to finance the sickness insurance premium for pensioners, their pension fund paid a lump sum to sickness insurance schemes. The contribution of the pension funds, however, was not sufficient to cover pensioners' health care costs. After the Second World War, in 1956, pensioners gained the same rights and benefits of the sickness insurance system as employed members (Förtsch, 1995: 267; Paulus, 1973). Finally, other segments of society, such as farmers (1972), the disabled (1974/1975), university students (1975), and artists and writers (1981) were required to join the statutory sickness insurance system. As the OECD Health Data show, from 1972 the statutory sickness insurance system covered about 90 per cent of the German population. The remaining 10 per cent were either covered by private sickness insurance and/or by government health care insurance for civil servants. Only about 1 per thousand did not have any sickness insurance coverage and therefore had to finance health care through their own means or, if this was not feasible, to seek public welfare assistance (*Sozialhilfe*) after falling ill (OECD, 1997; Bundesministerium für Gesundheit, 1997: 292–3).

How did relations between physicians, patients, pharmacists, and sickness insurance schemes change after 1883? In general, sickness insurance schemes contracted physicians to treat their members (Förtsch, 1995: 79–84). These physicians were chosen by the sickness insurance schemes and became so-called panel doctors (*Kassenärzte*). When a patient consulted a panel doctor, he or she had to present a treatment voucher of the insurance scheme so that the physician could claim his fee from that particular scheme. This meant that after 1883 the physician was no longer paid by the patient directly, but instead by the patient's sickness insurance scheme which also paid for medication.[2] Physicians decided on the therapy, which medication patients needed and whether hospital treatment was required. Physicians' decisions were founded on their medical expertise and not on economic considerations. Therefore it was the physician who determined the cost of treatment. Physicians were also, like all other health care providers, economically on the 'supply' side. This twofold function fundamentally changed the doctor-patient relationship in the course of the nineteenth century. As the gap in medical-scientific knowledge between patients and physicians widened, it became more and more difficult for patients to judge the efficiency of the treatment they received. The patient, suffering from illness, was put under pressure, having lastly only a limited choice when consulting a physician or healer. Eventually, the patient came to rely totally on the physician's medical expertise. This gap is decisive in understanding the situation at the end of the nineteenth century; it plays an even more important role in today's health care markets (Schluchter, 1980). While patients remained consumers, they lost their function as paymasters of health care. This function was assumed by the sickness insurance schemes which in economic terms represented 'demand'. On the other hand, sickness insurance schemes were not in a position to influence physicians' decisions on treatment and thus on costs.

In most markets one and the same person acts in three different roles. That person determines what to consume, and is consumer as well as paymaster, hence being able to decide what goods or services are considered essential and affordable. In the German health care market, after the 1883 Sickness Insurance Act, these three roles were allotted to different agents, namely the physician, the patient and the sickness insurance scheme. Furthermore, in contrast to 'normal' markets, supply and demand on the health care market were not regulated by market conditions but largely negotiated between suppliers and their organisations on the one hand and sickness insurance schemes and their affiliated organisations on the other (Herder-Dorneich, 1994).

The Monetary Side of the German Health Care System

Premiums provided the main income for sickness insurance schemes. Employees paid two-thirds and employers paid one-third (since 1951 both pay an equal share). The amount of the premium, however, was not determined by the worker's risk of falling ill, which would have been the principle used by an insurance company, but by gross income level. Those who earned less paid a lower premium than those who earned more. Despite the difference in payments, both were entitled to the same sickness insurance benefits in kind. This became the central principle on which compulsory sickness insurance operated, in effect reallocating resources within society. In 1913, workers paid an average premium of 1.9 per cent of their income to their sickness insurance scheme; in the late 1980s, the premium reached a level of approximately 13 per cent. Before the First World War, the burden of the sickness insurance premium was rather modest for both workers and their employers. Today, the premium has reached a level that is felt to be a burden on both employees and employers (Reuter, 1980: 126; Alber, 1992: 41), a result of permanently rising expenditure for sickness insurance schemes.

Sickness insurance expenditure (excluding administration) rose from 47 million Marks in 1885 to 445 million Marks in 1914, to 909 million Reichsmarks in 1925, and reached its peak at the beginning of the world economic crisis in 1929 with 1,862 million Marks. By 1933 expenditure had been reduced to 966 million Marks. While in 1885 a typical sickness insurance scheme spent 11 Marks per member, this amount rose to 29 Marks in 1914, 50 Marks in 1925, and 89 Marks in 1929 and was then reduced to 56 Marks in 1933 (Statistisches Bundesamt, 1972: 219). After the Second World War, sickness insurance expenditure rose again. In 1960 sickness insurance schemes spent 331 Marks per member and in 1990 3,538 Marks in Western Germany alone (Bundesminister für Gesundheit, 1997: 322). The share of sickness insurance expenditure of the GNP rose from about 0.2 per cent at the turn of the century to more than 6 per cent in the 1980s (Alber, 1992: 41).

In the decades before the 1970s, expansion of the health care sector was discussed under the expectation of being socially progressive. However, world-wide economic recession at the beginning of the 1970s led to a fundamental change in the perception of the sickness insurance system in Germany. On the one hand, health care expenditure increased considerably, from 4.3 per cent GDP in 1960 up to 8.0 per cent in 1975 (OECD, 1997). On the other hand, due to the recession at the beginning of the 1970s, the increasing unemployment rate threatened the income of the statutory sickness insurance schemes. Consequently, sickness insurance funds had to raise their premium from an average of 8.1 per cent in 1970 to 11.3 per cent in 1976. This measure added further burdens not only on the insured but on the country's economy which was already in recession. 'Cost explosion' became a key phrase in health care policy debates and the focus moved to discussing the economic aspects of health care. Among the first to introduce the problem of a 'cost explosion' in political terms was Heiner Geißler, then Minister of Social Affairs in the Rhineland-Palatinate (Kühn, 1995; Herder-Dorneich, 1994: 209–33).

There were several reasons for this enormous increase in health care expenditure. During the first decades of its existence, the major cause was a steady extension of benefits, both those voluntarily introduced by sickness insurance schemes and those imposed by state legislation. Due to the expansion of medical research, diagnostic and therapeutic measures became more costly while the demand for health care increased. Moreover, the proportion of people over 60 increased (and is still increasing), a development adding a further burden to health care costs, as the last year of life is, from a medical point of view, the most expensive one. This situation became even more difficult as the number of elderly people rose while those in productive work declined. This meant those paying into the statutory sickness insurance system, on whom health care systems depend financially, decreased.

Benefits of Sickness Insurance Schemes

Initially, sickness insurance schemes were legally obliged to pay for medical treatment and medication for 13 weeks from the first day of sickness onwards. In 1904, this period was extended to 26 weeks. Eyeglasses and trusses were financed as well as a funeral allowance, and women in childbed received a childbed allowance of up to a maximum four weeks after labour. It was the policy of many sickness insurance funds to expand the benefits for their members. This was possible because the 1883 Sickness Insurance Act and subsequent amendments left enough leeway for individual decisions within the insurance schemes. Often this policy met with resistance from employers who were afraid of a possible increase in insurance premiums because of these increasing benefits. Hence sickness insurance schemes offered their members different benefits within certain margins.

The government standardised the benefits of the sickness insurance system to some degree in the 1911 Imperial Insurance Act (*Reichsversicherungsordnung*). Under this legislation and again in 1920 under special legislation, the benefit for women in childbed was extended to eight weeks. Sickness insurance funds were also allowed to spend money on prophylactic medicine. But as a consequence of state legislation, sickness insurance schemes lost their competence to decide on their benefits (Förtsch, 1995: 160–61, 175, 208–9; Rother, 1994). During the inflation of 1922–1923 they were forced to slash their benefits drastically. Insurance scheme members were obliged to contribute to the costs of medication and dressings. Because of rising costs sickness insurance funds tried to reduce their expenditure and were successful in doing so by selling drugs through their own pharmacies. In the decade after 1923, sickness insurance benefits were expanded again (Förtsch, 1995: 176–85).

After the Second World War and up to the mid-1970s, the benefits of the sickness insurance system were once more extended. From 1971 early detection of diseases that would physically or mentally endanger the development of children under the age of four years and medical check-ups for detecting cancer in adults were financed by sickness insurance funds on a regular basis. Since 1974 all of those insured have been legally entitled to hospital treatment without a time limit. Previously sickness insurance schemes were required in advance to agree to a schedule in paying for the costs and would only do so for a period not longer than a consecutive 78 weeks (Tennstedt, 1976: 421).

In the mid-1970s perception of the expanding sickness insurance system changed. Whereas in the past additional benefits provided by sickness insurance schemes were seen as social progress, this changed under the influence of recession and the oil-crisis. Now cost-explosion became a key-term in health politics. The question was how to contain costs and still provide high level health care for all citizens in need. Several cost-containment laws have been enacted since then. Besides restrictions of benefits in kind the most contested issue was to which degree patients should contribute financially towards their health care costs, especially in regard to dentures.

Besides benefits in kind, sickness insurance schemes also provided cash benefits. Roughly up until the mid-1950s cash benefits exceeded benefits in kind. If the patient was unfit for work, the sickness insurance paid sickness benefits (*Krankengeld*) after a waiting period of three days (*Karenztage*) and lasting up to 13 weeks of sickness to support the worker and the worker's family. In 1904, this period was extended to 26 weeks. Although sickness benefits did not exceed half of the worker's average wage, they did improve his and his family's social situation immensely, since before the Sickness

Insurance Act of 1883, workers had not received any pay when they were too sick to work. At that time, sickness meant rapidly sinking into poverty and social distress. The main purpose of the statutory sickness insurance system was to preclude this slide into poverty should a family's breadwinner be too ill to work. Hence, in the 1890s, the ratio between cash benefits and benefits in kind was 1.7 to 1 (Tennstedt, 1976; Alber, 1992: 41).

In contrast to blue-collar workers, white-collar workers continued to receive their salary from their employer during the first six weeks of illness without any reduction. This was guaranteed through emergency legislation in 1930 and 1931 to reduce expenditure during the World Economic Crisis. This inequality between blue-collar workers and salaried employees was removed on an incremental basis after the Second World War. In particular, the trade unions demanded that wages continue to be paid in full to sick workers. They were successful in 1961 when employers had to assume the remaining share of blue-collar workers' wages not covered by the sickness insurance. Moreover, the waiting period after which blue-collar workers were entitled to sickness benefits was reduced to one day. Even though financial equality between blue-collar and white-collar workers was gradually assured, statutory differences remained. Sickness benefits for blue-collar workers were paid in part by their sickness insurance funds and in part by their employer, whereas sickness benefits for white-collar workers were solely paid by the employer. Thus, blue-collar workers financed a share of their sickness benefits through their insurance premium whereas this was not the case for white-collar workers. This discrepancy was finally removed in 1969 when – against fierce resistance by the Liberal Party and the Organisation of Employers (*Bund Deutscher Arbeitgeber*) – the 'Grand Coalition' of SPD and CDU/CSU agreed on the *Lohnfortzahlungsgesetz* which guaranteed sick blue-collar workers their full income from their employers for a maximum of six weeks (Reucher, 1999: 86–98, 158–60 and 225–6; Immergut, 1986; Webber, 1988: 188–9). By now, benefits in kind exceeded cash benefits by a factor of ten.

At the end of the nineteenth century, workers became members of the statutory sickness insurance system and thus were entitled to professional medical treatment and sick benefits financed by their insurance scheme. They obtained this right because they paid a certain percentage of their wages as a premium towards their sickness insurance scheme. This insurance premium was shared between employees and their employers. Until 1951 employers paid one-third of this premium, since then both employees and employers have paid an equal share. The amount of the premium depended on the workers' income and not on their individual risk, as is the case with private sickness insurance companies. All those insured were entitled to the same benefits in kind provided by their sickness insurance scheme. Hence Germany's statutory sickness insurance system became the most far-reaching institution for income redistribution within the social security system. This was based on a fundamental principle of solidarity among insurees which included the sick and the healthy, regardless of their income, childless singles as well as married couples with children, and all others, regardless of their sex and age (Ullrich, 1996).

Changes in the Health Care Market

An important consequence of the Sickness Insurance Act of 1883 was an increasing demand for medical services resulting in an enormous expansion of the health care market. Firstly, the number of physicians (including surgeons) multiplied. In 1876, eight years before the Sickness Insurance Act, there were 13,700 doctors in Germany – physicians as well as surgeons. This is an average of 3.2 doctors per 10,000 inhabitants. In 1900, the number of doctors had doubled (27,000). Now, there was an average of 4.9 doctors per 10,000 inhabitants. Only five years later, there were an additional 4,000 doctors. The ratio of doctors increased from 4.9 doctors per 10,000 inhabitants in 1900, to 5.1 in 1905, to 7.5 in 1930, 14.3 in 1960, 16.4 in 1970 and 33.6 in 1995. In 1995 more than 270,000 doctors were registered in Germany. The number and ratio of dentists had increased as well, though on a much lower level. In 1960 32,000 dentists were registered in Germany. Their number nearly doubled by 1995 (60,600). The ratio of dentists per 10,000 inhabitants increased from 5.9 in 1960 to 7.4 in 1995. Of course, there are other reasons for an increase in the number of physicians. For example, during the two World Wars, there was an increasing need for doctors to serve either at home or at the front. Under war conditions even doctors with poor qualifications passed their state exams (Wolff, 1997: 106–7).

A similarly impressive development can be observed with regard to hospitals and their bed capacity. In 1877, there were about 2,400 hospitals in Germany (Spree, 1995). But more important than the actual number of hospitals was the number of hospital beds available per 10,000 inhabitants. In 1877, there were 24.6 hospital beds per 10,000 inhabitants; by 1901 this figure had nearly doubled (48.3 hospital beds per 10,000). In 1930 this figure had nearly doubled again, so that now there were 90.9 hospital beds available for every 10,000 inhabitants. In the mid-1970s a peak of about 118 hospital beds for every 10,000 inhabitants was reached which has since been reduced to 97 (OECD, 1997). Moreover, not only did the number of hospitals and hospital beds increase, but so did the number of patients treated in these hospitals. Whereas in 1877 108.3 patients per 10,000 inhabitants were treated in a hospital, this figure doubled by 1901 (255.3) and nearly tripled by 1930 when 656.6 patients per 10,000 inhabitants received hospital treatment. In 1970 as many as 1,539.6 patients per 10,000 inhabitants were hospitalised.

Observing the expanding health care market in the late nineteenth century, the chemical industry in Germany realised that there was a potential market opening for them as well (Sauerteig, 1996; Wimmer, 1994). Until the second half of the nineteenth century, the production of therapeutic drugs was the domain of the pharmacies. This began to change in the mid-nineteenth century. From 1880, not only were more and more drugs being produced, but new types of drugs were being developed by the pharmaceutical industry. There were two important prerequisites for this: the new technical and chemical procedures which made the mass-production of drugs possible and the existence of a new and expanding health care market where more drugs could be sold. The role of the pharmacists changed in response to these developments. Whereas at the beginning of the nineteenth century pharmacists were responsible for the production of nearly all available drugs, at the beginning of the twentieth century their role was more or less reduced to selling drugs produced by pharmaceutical companies such as Merck, Schering, Hoechst or Ciba, to mention just the larger ones. Although the number of pharmacists increased in the nineteenth and early twentieth centuries, there were never more than 1.5 pharmacists per 10,000 inhabitants. This only changed to a certain degree after the Second World War. In 1960 there were 2.9 pharmacists per 10,000 inhabitants and in 1990 there were 5.5.

Many more figures could be quoted for other segments of the health care market. The trend, however, is clear: the health care market in Germany expanded tremendously in the nineteenth century. This development was markedly accelerated by the introduction of the sickness insurance legislation of 1883. Since then, the health care market has exploded. Two factors are relevant in explaining this development. Firstly, there were pull factors such as the increasing capacity of medicine to provide successful treatment, especially in surgery. Secondly, there were push factors. Here, the sickness insurance legislation was definitely the most important factor. For the first time sickness insurance schemes offered an increasing number of lower and middle class patients the opportunity to seek professional medical treatment and medical services. As the central financier, the sickness insurance system was also responsible for restructuring the socio-political relations on the health care market.

Sickness Insurance Schemes and the Health Care Market

The sickness insurance schemes are now the main financing agents on the health care market. However, in the course of the twentieth century, the sickness insurance schemes suffered from dwindling influence and power. The history of the sickness insurance schemes is, one could say, a history of their increasing inability to control expenses (Förtsch, 1995: 160–62). Only recently has Germany's health care legislation stopped or even reversed this development by giving back to the insurance schemes some of the power they originally possessed.

The decline of the sickness insurance schemes' power was the result, on one level, of their changing relationship to the medical profession. In the first decades after the Sickness Insurance Act, insurance schemes contracted a certain number of physicians. There were different types of contracts: one or more panel doctors were employed by one or more insurance schemes to treat their members in a certain district (*Distriktarztsystem*), or some or all local physicians were allowed to treat members of an insurance scheme (*beschränkte freie oder freie Arztwahl*). These panel doctors were under the control of the sickness insurance schemes because they wanted to force the panel doctors to take into account the economic advantage of the insurance schemes. Thus, to some degree, the sickness insurance schemes influenced medical treatment. Members of insurance schemes were only allowed to seek treatment through panel doctors, and the patients' original freedom to choose any physician he or she trusted was strictly limited. From the 1890s opposition to these restrictions arose amongst physicians, especially amongst those who were excluded from the panel system. Physicians argued against exclusion on the basis of the patients' right to freedom of choice, but also to enhance their own professional interests (Herold-Schmidt, 1997: 84; Förtsch, 1995: 79, 83).

Consequently, the medical profession obtained a more tightly knit organisational structure. The most important organisation in the first decades of the twentieth century was the Association of Physicians of Germany for the Protection of their Economic Interests (*Verband der Aerzte Deutschlands zur Wahrung ihrer wirtschaftlichen Interessen*) which was founded in 1900 in Leipzig by the physician Hermann Hartmann (1863–1923) and thereafter called the *Hartmann Bund*. Before the First World War about two-thirds of all German physicians became members of the *Hartmann Bund*. Much like a trade union for physicians, the *Hartmann Bund* promoted the economic interest of physicians against the policies of sickness insurance schemes. The organisation improved their position relating to the sickness insurance schemes and did not hesitate to threaten strikes. Between 1900 and 1912 there were about 1,000 strikes organised by the *Hartmann Bund* (Herold-Schmidt, 1997: 50–51; Neuhaus, 1986; Huerkamp, 1985; Tennstedt, 1976: 394). In 1913, with the support of the regional governments, the medical profession reached an agreement with the sickness insurance schemes regulating relations for the next decade. One of the major points of this so-called Berlin agreement was that the sickness insurance schemes were forced to register at least one panel doctor for every 1,350 members. Which physician should be registered was decided by a special committee consisting of members from physicians' organisations and sickness insurance schemes. Furthermore, the sickness insurance schemes could no longer secure contracts with individual physicians, but had to agree to collective contracts with all panel doctors. In 1923, these regulations were included in the Imperial Insurance Act (Neuhaus, 1986; Tennstedt, 1976: 395–7).

During the inflation in 1923–1924 the economic situation of sickness insurance schemes deteriorated. A governmental regulation obliged physicians to take the economic situation of sickness insurance schemes into consideration when treating patients. Physicians however did not agree with this regulation and went on a general strike in November 1923. As a reaction some sickness insurance schemes, especially those in cities, opened dental clinics and so-called *Ambulatorien* (out-patient departments) employing their own physicians to treat insured members. Hence, sickness insurance schemes also tried to control their increasing expenses for physician's fees during the inflation. Physicians' organisations, especially the *Hartmann Bund*, vociferously protested against any form of treatment centres run by sickness insurance schemes (Förtsch, 1995: 191; Wolff, 1997: 114–9; Döhler, 1984; Hansen, 1981; Tennstedt, 1976: 398–9).

The next blow to the status of the sickness insurance system was dealt by the World Economic Crisis. The entire social security system, especially unemployment insurance, but also the sickness insurance schemes, was subject to a deep financial crisis. There were several reasons for this. Firstly, in the decade before this crisis, the sickness insurance schemes had considerably expanded benefits for their members, a policy which had, of course, also increased their expenses. Secondly, because of the rising unemployment rate, the insurance schemes lost many of their premium-paying members (some 5.7 million members by 1932). Furthermore, due to reduced wages, the revenues of the insurance schemes decreased even more. It became, therefore, extremely difficult for insurance schemes to balance their budget. A further increase of the insurance premium, which had already been raised several times since the end of the war, would not only have been a further burden for the members but also for the employers, who were already suffering greatly from the economic crisis.

In this situation the German Reichs-President Heinrich Brüning tried to stabilise the financial position of the sickness insurance schemes. In 1930, he introduced charges for treatment vouchers and prescriptions (decree under emergency legislation, 26 July 1930). The waiting period after which workers received sick benefits was prolonged to four days instead of the original three, and the sickness benefit rate was reduced from 75 to 50 per cent of the worker's wage. By reducing the expenses of health insurance schemes in this manner, the insurance premium could be fixed at 5 per cent of wages, thus reducing employers' costs as well. Secondly, the sickness insurance schemes were subjected to closer governmental scrutiny under the Insurance Supervisory Boards (*Oberversicherungsämter*). The Insurance Supervisory Boards not only had to approve any increase of the insurance premium but could also force the insurance schemes to modify their premium rate (decree, 5 June 1931).

Thus, the insurance schemes lost most of their financial independence. Furthermore, the 1930 decree forced insurance schemes to reduce their benefits to the legal minimum. Thirdly, the Insurance Supervisory Boards were authorised to reduce the number of panel doctors the sickness insurance schemes had to contract. In 1931 the proportion of members of sickness insurance schemes to physicians was fixed at a ratio of 600 to one physician (before, the ratio was 1,000 to one physician). Thus, more physicians could attain the status of a panel doctor, and this guaranteed income. To stabilise the financial position of sickness insurance schemes physicians had to agree to a reduction of their fees. Also, a mechanism was introduced to link treatment expenditure of sickness insurance schemes to their revenue. Thus further increases in treatment expenditure could only be undertaken if the revenue of the insurance scheme was raised as well. In addition the insurance schemes were legally obliged to engage special medical examiners (*Vertrauensärzte*) to control panel doctors, especially their unfit-for-work-certificates (Webber, 1988: 171–6; Kohlhausen, 1976; Tennstedt, 1976: 401).[3]

The foundation of the Panel Doctors' Association (*Kassenärztliche Vereinigung*) in 1931 fundamentally changed the relationship between insurance schemes and physicians (Wolff, 1997: 132–3; Webber, 1992). This Association became a Body of Public Law (*Körperschaft öffentlichen Rechts*) and every panel doctor was obliged to become a member of one of the branches of the Association. Instead of reimbursing physicians directly, sickness insurance schemes now paid a lump sum for every member treated by a panel doctor (*Kopfpauschale*) to the Panel Doctors' Association. The idea behind this reform of the remuneration of physicians was to make sure that during the economic crisis physicians could not enlarge their income beyond the economic capacity of insurance schemes by increasing their services. The Panel Doctors' Association distributed the money to the individual panel doctor and took responsibility for monitoring panel doctors so that they would treat their patients not only sufficiently and effectively, but also economically. Furthermore, the Association had to ensure the availability of a sufficient number of panel doctors, a task which the insurance schemes had themselves undertaken in the past. To comply with these duties the Association was granted disciplinary power over their members. As sickness insurance schemes lost their power over panel doctors they were deprived of yet another element of their financial sovereignty. Panel doctors became relatively independent of sickness insurance schemes (Förtsch, 1995: 235–40; Alber, 1992: 48–50; Sachße and Tennstedt, 1988).

Within the first years of the Nazi Period, the statutory sickness insurance system suffered a further decline in power. In 1933–1934 the NS regime abolished the principle of self-administration and put sickness insurance schemes under tighter governmental control. In 1937 the central organisations of the sickness insurance schemes became, like the Panel Doctors' Association, Bodies of Public Law. With the new legislation pertaining to Panel Doctors'

Legislation, the sickness insurance schemes completely lost their remaining influence over panel doctors. The registration of panel doctors became the sole responsibility of the Panel Doctors' Association. Furthermore, sickness insurance schemes had to close down their own out-patient departments (*Ambulatorien*). As a result, the position of the medical profession, amongst the first to welcome the NS regime (Kater, 1989), was strengthened, while the position of the sickness insurance schemes was weakened further. This development has to be seen against the backdrop of Nazi racial ideology with its fundamental change of focus from the individual to the national community (*Volksgemeinschaft*). The idea of solidarity was turned upside down. The duty of the community to care for the weak became the duty of everybody to subordinate his or her interest to the interest of the national state. As a consequence, the legal right to social services was abolished. It is not surprising that the sickness insurance schemes partly financed compulsory sterilisation performed on their members (Förtsch, 1995: 241–70; Sachße and Tennstedt, 1992; Alber, 1992: 51–5; Weindling, 1989; Webber, 1988: 177–82; Teppe, 1977; Tennstedt, 1977: 181–225; Tennstedt, 1976: 405–8).

After the Second World War, in the Allied Occupied Zones the sickness insurances' and doctors' associations returned to pre-1933 conditions. Self-administration was re-established within the social insurance system in 1951, albeit with important changes. The sickness insurance schemes, however, did not regain the power they had had at the beginning of the Weimar Republic. Before 1934, when self-administration was abolished by the Nazis, employees had a two-thirds majority of the votes in the administration of sickness insurance schemes. Against resistance from trade unions and Social Democrats, the Conservative government, with support from the Liberals and employers' organisations, decided in favour of self-administration with equal numbers of employers' and employees' representatives. While previously employees paid two-thirds of the insurance premium, they now only had to pay 50 per cent. Since then the question of self-administration has become politically insignificant (Webber, 1988: 182–5; Hockerts, 1980; Tennstedt, 1977: 243–61; Tennstedt, 1976: 414–5).

The Panel Doctors' Association was also re-established, and centralised on an incremental basis (*Kassenärztliche Bundesvereinigung*). With the 1955 Legislation on Panel Doctors (*Gesetz über Kassenarztrecht*), the Panel Doctors' Association was given the sole responsibility of providing physicians for all out-patient services in Germany (Döhler and Manow, 1997: 37–9). In return they had to guarantee that physicians would refrain from strikes of any kind in the future. Based on this legislation the Panel Doctors' Association gained an even more powerful and influential position on the health care market than in the Weimar Republic. It succeeded in securing an increasing income for panel doctors over the next decades. One factor for improving physicians' income in the short term was a change in the remuneration system. During the first half of the 1960s, Panel Doctors' Associations convinced sickness insurance schemes to reintroduce the fee-for-service system (Herder-Dorneich, 1994: 351–9).

Furthermore, in 1955, sickness insurance schemes had to agree to reducing the number of members per physician to 500. This measure led to an increase in the number of panel doctors and, thus, improved the population's access to medical services. Under this regulation the Panel Doctors' Association was able to regulate the distribution of physicians in Germany, thereby ensuring that there were enough physicians available in rural areas which generally lacked medical services. In 1960, however, the Federal Constitutional Court (*Bundesverfassungsgericht*) abolished all former restrictions limiting the number of panel doctors. Hence, every physician (and from 1961 onwards every dentist as well) had the right to be registered as a panel doctor. This was not only one of the major causes for the increase in costs of the sickness insurance schemes but also made it very difficult for the Panel Doctors' Association to ensure a sufficient and equal distribution of physicians throughout Germany, especially for rural areas (Gerst, 1997: 197–8, 210–12, 216–19, 222–4; Förtsch, 1995: 271–322; Alber, 1992: 55–62; Hockerts, 1983; Tennstedt, 1976: 415–7). Restrictions for the registration of panel doctors were not reintroduced until very recently.

There is not enough space here to analyse in greater detail the changing relationship of sickness insurance schemes to other agents on the health care market, such as hospitals or the pharmaceutical industry. However, the trend is similar. Originally, around the turn of the century, sickness insurance schemes were in quite a powerful position. They made the decision as to which hospital to accept for their members and they negotiated the hospital allowance which every single registered hospital received for its medical services (Förtsch, 1995: 99–107). As mentioned above, to control the price for therapeutic drugs, larger sickness insurance schemes founded their own pharmacies (Förtsch, 1995: 107–9; Hansen, 1981). But in the course of the twentieth century, the sickness insurance schemes became increasingly incapable of controlling the rocketing expenses for hospital care (Simon, 2000) and of medicine. Especially in the postwar period, pressure from the medical profession, the hospital administrations, and the pharmaceutical and medical-technological industries to acquire a larger proportion of the budget of sickness insurance schemes mounted steadily; however, reform plans to stop the increasing health care expenses in the late 1950s and early 1960s failed (Reucher, 1999; Döhler and Manow, 1997: 33–45).

The 1977 Sickness Insurance Cost Containment Act (Herder-Dorneich, 1994: 141–50) on the one hand imposed restrictions on sickness insurance schemes regarding their benefits in kind and on patients who now had to pay increasing prescription charges. On the other hand, a new institution to emerge on the health care market was the Concerted Action Conference for Health

(*Konzertierte Aktion im Gesundheitswesen*) which united key figures in health care policy, altogether more than 60 members from the central organisations of the different sickness insurance schemes, private sickness insurance companies, panel doctors and dentists, the Federal Doctors' Organisation (*Bundesärztekammer*), hospitals, the pharmaceutical industry, pharmacists, trade unions and employers, and the government at its various levels, i.e. local, federal and central government. The main functions of the Concerted Action Conference were, firstly, to provide data pertaining to medical and economic development on which all participating groups could agree and, secondly, to suggest recommendations for a more rational, efficient, and effective health care system. This was to be achieved on the basis of an adequate health care system at the level of current medical knowledge for all citizens and a balanced distribution of financial burdens. The overall idea was to achieve an income-oriented expenditure policy (*einnahmenorientierte Ausgabenpolitik*). The revenue of the sickness insurance system should define the frame for expenditure and thus for benefits as well as physicians' fees etc. As health economists argued, the Concerted Action Conference, lacking the power to impose any binding regulations, was only an instrument of moral persuasion of those involved in providing health care, urging them to take the general economic limitations in health care policy into consideration and, thus, to make compromises more feasible (Döhler and Manow, 1997: 94–7; Herder-Dorneich, 1994: 395–423; Wiesenthal, 1981).

This aim was largely achieved during the first years of the Concerted Action Conference. The increased costs of out-patient health care were reduced, as well as costs for medication. However, as hospital organisations did not participate, costs for hospital treatment exploded even further and reached hitherto unforeseen proportions. Health insurance schemes were largely deprived of the power to control their costs. As the debate on the Cost Containment Act shows, government intervention increased through establishing measures for cost containment from the mid-1970s onwards. However, against stiff resistance from the different interest groups involved in the health care market (panel doctors, hospitals, pharmaceutical industry), it was difficult to find majorities for more far-reaching changes in health care politics. On the other hand, under the pressure of the economic situation in general, questions regarding the just allocation of resources in health care became increasingly important. Since the mid-1970s, one could argue, the debate on health care policy was dominated by economic aspects (Herder-Dorneich,1994:185–6).

The debate on cost containment in health care in the 1970s brought far-reaching changes for panel doctors. The Cost Containment Act of 1977 specified that Panel Doctors' Associations must take the economic capacity of sickness insurance schemes, indicated by the level of average salaries, into account when negotiating their contracts. This Act also introduced a cost level which physicians were not allowed to exceed when prescribing medication (*Arzneimittelhöchstbetrag*). If a physician prescribed medication above this ceiling, the sickness insurance scheme could claim money back. Patients were required to pay higher health care charges. The SPD/FDP Coalition, however, did not succeed in implementing measures to control hospital costs (Simon, 2000). Any such policy was thwarted by a majority of the federal states in the *Bundesrat*. Nevertheless, the cost containment policy was successful in averting a cost explosion in health care up until the beginning of the 1980s. The proportion of the GDP spent on health care in total remained constant, even dropping in the period between 1975 and 1981 (from 8.0 per cent in 1975 to 7.8 per cent in 1977) (OECD, 1997). The situation changed again with a deteriorating economy at the beginning of the 1980s (Webber, 1988: 193–9).

When the Conservative-Liberal Coalition took over in 1982, they continued the cost containment policy of the former coalition government of Social Democrats and Liberals. Like the former government, the Conservative-Liberal Coalition aimed at stabilising the premium rate for the sickness insurance system and therefore reducing an increase in social security costs. Since 1970 the average premium for statutory sickness insurance had increased from 8.2 per cent to 12 per cent in 1982. The key idea behind this policy was to make German enterprise internationally competitive by reducing incidental labour costs (*Lohnnebenkosten*). Whereas the average sickness insurance premium dropped during the following two years (11.44 per cent in 1984), the problem of cost containment was put back on the political agenda in the second half of the 1980s in conjunction with a renewed increase in the average premium rate (12.9 per cent by 1988).

At this juncture the Coalition discussed more fundamental and structural changes in the health insurance system. One of the major aims was to fight what Conservatives and Liberals called the 'claim mentality' (*Anspruchsmentalität*) or 'free beer mentality' of members of the social security system who always expected services in situations when they could have helped themselves or who even abused services (moral hazard problem). Although the Coalition succeeded in passing health care reform legislation in 1988 (*Gesundheitsreformgesetz*), they failed to implement any substantial structural reforms. In general patients had to pay health care charges while the position of physicians and hospitals remained largely untouched. Sickness insurance schemes gained more control over the price of medication. Once more patients had to bear an exceptionally large share of their health care costs, especially in dental care. By now the proportion of health care costs directly financed by patients had increased from 2.7 per cent of health care expenditure in the statutory sickness insurance system in 1977 to approximately 8.5 per cent (Blanke and Kania, 1996: 526–9; Ullrich, 1995; Perschke-Hartmann, 1994; Pfaff, Busch and Rindsfüßer, 1994; Döhler, 1990: 409–502; Webber, 1989).

However, this legislation did not effectively limit health care costs. The statutory sickness insurance system was running into deficit again and an increase in the insurance premium was imminent. Thus, cost limitation remained on the political agenda during a period of economic downturn and political pressure. The result was a new health care reform act (*Gesundheitsstruktur-gesetz*) passed in 1992 which fundamentally changed the relationship between sickness insurance schemes, health care providers and the government. Although the principle of self-administration was not abolished, the government gained a more influential position in controlling health care provision and its economic structure and assumed more responsibility. The state intervened in regulating prices by allocating fixed budgets to all sectors of health care services (hospitals, physicians and dentists, and medication). Any changes in these expenditure ceilings were tied to an increase in basic salaries and hence of the revenue of the sickness insurance system. Furthermore, the pharmaceutical industry was forced by law to reduce the prices it charged for its products by 5 per cent. The government implemented control strategies to intervene in the self-administration and self-regulation of health care providers and sickness insurance schemes in order to achieve overall cost control in health care. However, this reform also introduced initial measures for a more market orientated health care policy by giving more freedom in decision-making to sickness insurance schemes. Now insurance schemes were able to offer special arrangements for their members and gained power in dealing with health care providers. It was hoped that through such a policy, competition amongst the different sickness insurance schemes would be increased (Blanke and Perschke-Hartmann, 1994; Pfaff, Busch and Rindsfüßer, 1994; Perschke-Hartmann, 1993).

Conclusions

For more than a century, the sickness insurance schemes were the major financing agents on Germany's health care market. From the start, they were the driving force behind its enormous expansion. This development resulted from the fact that the sickness insurance system provided access to health care services for many levels of society that had little access to professional medical treatment before this. Hence, the sickness insurance system was the main factor in achieving equality of access to health care in Germany. Equality in the allocation of resources was and still is one of the fundamental principles of Germany's health care policy extending from the Weimar Republic to the Federal Republic.[4]

This chapter has shown that the prime mover in the health care market, with its role of 'buyer' and representing the 'demand function' on the market, i.e. the sickness insurance schemes, was gradually deprived of most of its power to control expenses. As a consequence of continually increasing demands for the best medical treatment available, the latest diagnostic technology, and the most modern hospital care, the share of Germany's GDP spent on health care increased, although several reforms tried to halt this process. However, what was perceived as being a cost explosion in health care since the mid-1970s was really not so much an explosion of costs but rather a drop in the revenue of sickness insurance schemes (Kühn, 1995).

In absolute figures, health care costs did increase but not at a much faster rate than Germany's GDP. There seems to be a general unwillingness to spend more money on health care since the mid-1970s when the key phrase 'cost explosion' was introduced to policy-making in health care. While in the past health care policy largely concentrated on expanding the social security system, bringing a wider range of citizens under statutory sickness insurance coverage, and offering better health care services financed by the sickness insurance system, this changed during the first half of the 1970s. Under the influence of the economic recession, increasing health care expenses were perceived as a 'cost explosion'. While previously health care policy had been dominated by the ideal of social progress, it was now dominated by an economic discourse. As a consequence of this 'economisation' of the debate on health care, premiums for the sickness insurance system – like those for other social insurance benefits – were increasingly seen as an economic burden that undermined economic stability and progress. The question the changing governments in Germany had to solve concentrated increasingly on the proportion of the GDP that should be spent on health care (Blanke and Kania, 1996), and who should finance health care costs. The overall trend since the second half of the 1970s has been to increase the financial burden of patients, thus undermining the principle of solidarity. One could therefore speak of a slow privatisation of health care costs.

Both Germany's and Britain's health care systems, although financed differently, aim at equity in the allocation of resources to all citizens. In Britain, however, the state as the financier of the NHS succeeded to a much greater degree in controlling expenses in health care. The proportion of the British GDP spent on health care has increased only slightly since the 1970s (4.5 per cent in 1970, 5.6 in 1980, 6.0 in 1990), and in 1996, at 6.9 per cent, reached a level that Germany had already exceeded at the beginning of the 1970s. In Germany the proportion of the GDP spent on health care nearly doubled during the same period from 5.7 per cent in 1970 to 10.5 per cent in 1996 (OECD, 1997).

What this meant in each country becomes clear when comparing the total per capita expenditure on health care under purchasing-power parity. In Germany, health care expenditure increased between 1960 and 1970 by a factor of 2.5 (from 91 $ to 230 $); in Britain it rose only by a factor of 1.9 (from 77 $ to 147 $). In the following decade, per capita health care expenditures increased slightly faster in both countries. But again Germany was ahead by a factor of 3.7 (to 860 $ in 1980), whereas British expenditure increased only by a factor of 3.0 (to 453 $ in 1980). The consequences of Germany's cost containment policy during the 1980s are evident from a slower increase of per capita health care expenditure (1,642 $ in 1990 and an increase by a factor of 1.9) than in Britain (957 $ in 1990 and an increase by a factor of 2.1). Until 1996, per capita health care expenditure increased even more slowly by the factor of 1.4 in both countries (2,222 $ in Germany and 1,304 $ in Britain in 1996). However, Germany still spent 1.7 as much per capita on health care as Britain and this difference has increased since 1960 when Germany spent only 1.2 times as much as Britain (OECD, 1997; Webster, 1988, 1996).

Recently it has been pointed out in a newspaper article that although Britain spends a lower proportion of its GDP on health care, life expectancy in both countries does not differ considerably and is even higher in Britain (Zitzelberger, 1998; Rubner, 1998). On this basis it seems that there is no longer a direct connection between mortality rate and health care expenditure in developed countries in Western Europe. This applies even more powerfully to the United States, which does not offer equal access to health care. This is, of course, a rather general statement which neglects the many existing differences in the quality of health care services between Germany, Britain and the United States. However, if one accepts this statement for the moment, then one could argue that the German health care system lacks equality because it deprives the working population of an unnecessarily large proportion of its income for a financially inadequate health care system.

Notes

1 Reuter refers to a definition of social policy given by one of the leading German economists of the nineteenth century, Adolph Wagner, in 1891.

2 The doctor-patient relationship in the consultation room might have changed as well – at least from the physicians' perspective, who now felt they were considered employees of their patients and reduced to mere writers of prescriptions, cf. Herold-Schmidt, 1997, pp. 85f.

3 Some sickness insurance schemes already had medical examiners, but now all of them were legally obliged to have them.

4 See § 161 of the Weimar Constitution: 'Zur Erhaltung der Gesundheit und Arbeitsfähigkeit, zum Schutz der Mutterschaft und zur Versorgung gegen die wirtschaftlichen Folgen von Alter, Schwäche und Wechselfällen des Lebens schafft das Reich ein umfassendes Versicherungswesen unter maßgebender Mitwirkung der Versicherten.'

References

Alber, J. (1992), 'Bundesrepublik Deutschland', in Alber, J. and Bernardi-Schenkluhn, B. (eds), *Westeuropäische Gesundheitssysteme im Vergleich. Bundesrepublik Deutschland, Schweiz, Frankreich, Italien, Großbritannien*, Campus, Frankfurt/M., New York, pp. 31–176.

Anderson, O.W. (1972), *Health Care: Can There Be Equity? The United States, Sweden, and England*, John Wiley, New York.

Bauer, G. (1976), 'Die Finanzwirtschaft in der Krankenversicherung', in Blohmke (1976), pp. 492–515.

Blanke, B. and Kania, H. (1996), 'Die Ökonomisierung der Gesundheitspolitik. Von der Globalsteuerung zum Wettbewerbskonzept im Gesundheitswesen', *Leviathan. Zeitschrift für Sozialwissenschaft*, vol. 24, pp. 512–38.

Blanke, B. and Perschke-Hartmann, C. (1994), 'The 1992 Health Reform. Victory over Pressure Group Politics', *German Politics*, vol. 3, pp. 233–48.

Blohmke, M. et al. (eds) (1976), *Handbuch der Sozialmedizin*, vol. 3, Ferdinand Enke, Stuttgart.

Bundesminister für Gesundheit (ed.) (1997), *Daten des Gesundheitswesens. Ausgabe 1997* (Schriftenreihe des Bundesministeriums für Gesundheit, vol. 91), Nomos Verlagsgesellschaft, Baden-Baden.

Döhler, M. (1984), 'Zur Entwicklung und Funktion der Eigeneinrichtung der Krankenkassen 1900–1933', *Zeitschrift für Sozialreform*, vol. 30, pp. 214–35 and 354–66.

Döhler, M. (1990), *Gesundheitspolitik nach der 'Wende'. Policy-Netzwerke und ordnungspolitischer Strategiewechsel in Großbritannien, den USA und der Bundesrepublik Deutschland*, Edition Sigma Bohn, Berlin.

Döhler, M. and Manow, P. (1997), *Strukturbildung von Politikfeldern. Das Beispiel bundesdeutscher Gesundheitspolitik seit den fünfziger Jahren*, Leske and Budrich, Opladen.

Förtsch, F. (1995), *Gesundheit, Krankheit, Selbstverwaltung. Geschichte der Allgemeinen Ortskrankenkassen im Landkreis Schwäbisch Hall 1884–1973* (Forschungen aus Württembergisch Franken, vol. 43), Jan Thorbecke, Sigmaringen.

Frevert, U. (1981), 'Arbeiterkrankheit und Arbeiterkrankenkassen im Industrialisierungsprozeß Preußens (1840–1870)', in Conze, W. (ed.), *Arbeiterexistenz im 19. Jahrhundert*, Klett-Cotta, Stuttgart, pp. 293–319.

Gerst, T. (1997), 'Neuaufbau und Konsolidierung. Ärztliche Selbstverwaltung und Interessenvertretung in den drei Westzonen und der Bundesrepublik Deutschland 1945–1995', in Jütte (1997), pp. 195–242.

Hansen, E. et al. (1981), *Seit über einem Jahrhundert ... Verschüttete Alternativen in der Sozialpolitik. Sozialer Fortschritt, organisierte Dienstleistungsmacht und nationalsozialistische Machtergreifung. Der Fall der Ambulatorien in den Unterweserstädten und Berlin*, Bund-Verlag, Köln.

Hentschel, V. (1983), *Geschichte der deutschen Sozialpolitik (1880-1980). Soziale Sicherung und kollektives Arbeitsrecht*, Suhrkamp, Frankfurt/M.

Herder-Dorneich, P. (1994), *Ökonomische Theorie des Gesundheitswesens. Problemgeschichte, Problembereiche, Theoretische Grundlagen*, Nomos, Baden-Baden.

Herold-Schmidt, H. (1997), 'Ärztliche Interessenvertretung im Kaiserreich 1871–1914', in Jütte (1997), pp. 43–95.

Hockerts, H.G. (1980), *Sozialpolitische Entscheidungen im Nachkriegsdeutschland. Alliierte und deutsche Sozialversicherungspolitik 1945 bis 1957*, Klett-Cotta, Stuttgart.

Hockerts, H.G. (1983), 'Hundert Jahre Sozialversicherung in Deutschland. Ein Bericht über die neuere Forschung', *Historische Zeitschrift*, vol. 237, pp. 361–84.

Huerkamp, C. (1985), *Der Aufstieg der Ärzte im 19. Jahrhundert. Vom gelehrten Stand zum professionellen Experten: Das Beispiel Preußens*, Vandenhoeck and Ruprecht, Göttingen.

Immergut, E. (1987), 'Between State and Market: Sickness Benefits and Social Control', in Rein, M. and Rainwater, L. (eds), *Public/Private Interplay in Social Protection. A Comparative Study*, M.E. Sharpe, Armonk, New York, London, pp. 57–98.

Jütte, R. (ed.) (1997), *Geschichte der deutschen Ärzteschaft. Organisierte Berufs- und Gesundheitspolitik im 19. und 20. Jahrhundert*, Deutscher Ärzte-Verlag, Cologne.

Kater, M.H. (1989), *Doctors Under Hitler*, University of North Carolina Press, Chapel Hill, London.

Kohlhausen, K. (1976), 'Vertrauensärztlicher Dienst', in Blohmke (1976), pp. 558–73.

Kühn, H. (1995), 'Zwanzig Jahre "Kostenexplosion". Anmerkungen zur Makroökonomie einer Gesundheitsreform', *Jahrbuch für kritische Medizin*, vol. 24, pp. 145–61.

Loetz, F. (1993), *Vom Kranken zum Patienten. 'Medikalisierung' und medizinische Vergesellschaftung am Beispiel Badens 1750–1850*, Franz Steiner, Stuttgart.

Neuhaus, R. (1986), *Arbeitskämpfe, Ärztestreiks, Sozialreformer. Sozialpolitische Konfliktregelung 1900 bis 1914*, Duncker and Humblot, Berlin.

OECD (1997), *OECD Health Data 97*, OECD Publication Service, Paris.

Paulus, K.H. (1973), *Die Krankenversicherung der Rentner (KvdR). Entwicklung und Reformen unter besonderer Berücksichtigung der Finanzierung*, Thesis, University of Cologne.

Perschke-Hartmann, C. (1993), 'Das Gesundheits-Strukturgesetz von 1992. Zur Selbstevaluation staatlicher Politik', *Leviathan. Zeitschrift für Sozialwissenschaft*, vol. 21, pp. 564–83.

Perschke-Hartmann, C. (1994), *Die doppelte Reform. Gesundheitspolitik von Blüm zu Seehofer*, Leske and Budrich, Opladen.

Pfaff, A.B., Busch, S. and Rindsfüßer, C. (1994), *Kostendämpfung in der gesetzlichen Krankenversicherung. Auswirkungen der Reformgesetzgebung 1989 und 1993 auf die Versicherten*, Campus-Verlag, Frankfurt/M.

Prange, P. (1954), 'Die gesetzliche Krankenversicherung in der Zeit der Weimarer Republik (1919 bis 1932)', in Rohrbeck, W. (ed.), *Beiträge zur Sozialversicherung. Festgabe für Dr. Johannes Krohn zum 70. Geburtstag*, Duncker and Humblot, Berlin, pp. 209–30.

Reucher, U. (1999), *Reformen und Reformversuche in der gesetzlichen Krankenversicherung (1956–1965). Ein Beitrag zur Geschichte bundesdeutscher Sozialpolitik*, Droste, Düsseldorf.

Reuter, H.-G. (1980), 'Verteilungs- und Umverteilungseffekte der Sozialversicherungsgesetzgebung im Kaiserreich', in Blaich, F. (ed.), *Staatliche Umverteilungspolitik in historischer Perspektive. Beiträge zur Entwicklung des Staatsinterventionismus in Deutschland und Österreich* (Schriften des Vereins für Sozialpolitik, vol. 109), Duncker and Humblot, Berlin, pp. 107–63.

Ritter, G.A. (1983), *Sozialversicherung in Deutschland und England im Vergleich*, Beck, Munich.

Ritter, G.A. (1991), *Der Sozialstaat. Entstehung und Entwicklung im internationalen Vergleich*, Oldenbourg, Munich.

Rother, K. (1994), *Die Reichsversicherungsordnung 1911. Das Ringen um die letzte große Arbeiterversicherungsgesetzgebung des Kaiserreichs unter besonderer Berücksichtigung der Rolle der Sozialdemokratie*, Verlag Mainz, Aachen.

Rubner, J. (1998), 'Wieviel Medizin können wir uns leisten? Von den Allmachtsphantasien endlosen Lebens und ihren Kosten für die Gesellschaft', *Süddeutsche Zeitung*, no. 216, 19/20 September 1998, supplement 'SZ am Wochenende', p. vi.

Sachße, C. and Tennstedt, F. (1980), *Geschichte der Armenfürsorge in Deutschland*, vol. 1: *Vom Spätmittelalter bis zum Ersten Weltkrieg*, Kohlhammer, Stuttgart.

Sachße, C. and Tennstedt, F. (1988), *Geschichte der Armenfürsorge in Deutschland*, vol. 2: *Fürsorge und Wohlfahrtspflege 1871 bis 1929*, Kohlhammer, Stuttgart.

Sachße, C. and Tennstedt, F. (1992), *Geschichte der Armenfürsorge in Deutschland*, vol. 3: *Der Wohlfahrtsstaat im Nationalsozialismus*, Kohlhammer, Stuttgart.

Sauerteig, L. (1996), 'Die Eroberung des Gesundheitsmarktes. Pharmazeutische Industrie und Gesundheitswesen um die Jahrhundertwende', *Wirtschaft und Gesellschaft*, issue 4, pp. 35–42.

Schluchter, W. (1980), 'Legitimationsprobleme der Medizin', in Schluchter, W. (ed.), *Rationalismus der Weltbeherrschung. Studien zu Max Weber*, Suhrkamp, Frankfurt/M., pp. 185–207 and 287–303.

Simon, M. (2000), *Krankenhauspolitik in der Bundesrepublik Deutschland. Historische Entwicklung und Probleme der politischen Steuerung stationärer Krankenversorgung*, Westdeutscher Verlag, Opladen.

Spree, R. (1995), 'Krankenhausentwicklung und Sozialpolitik in Deutschland während des 19. Jahrhunderts', *Historische Zeitschrift*, vol. 260, pp. 75–105.

Statistisches Bundesamt (ed.) (1972), *Bevölkerung und Wirtschaft, 1872–1972*, W. Kohlhammer, Stuttgart, Mainz.

Stollberg, G. (1983), 'Die gewerkschaftsnahen zentralisierten Hilfskassen im Deutschen Kaiserreich', *Zeitschrift für Sozialreform*, vol. 29, pp. 339–69.

Tennstedt, F. (1976), 'Sozialgeschichte der Sozialversicherung', in Blohmke (1976), pp. 385–492.

Tennstedt, F. (1977), *Geschichte der Selbstverwaltung in der Krankenversicherung von der Mitte des 19. Jahrhunderts bis zur Gründung der Bundesrepublik Deutschland* (Soziale Selbstverwaltung, vol. 2), Verlag der Ortskrankenkassen, Bonn.

Tennstedt, F. (1983), 'Die Errichtung von Krankenkassen in deutschen Städten nach dem Gesetz betr. die Krankenversicherung der Arbeiter vom 15. Juni 1883. Ein Beitrag zur Frühgeschichte der gesetzlichen Krankenversicherung in Deutschland', *Zeitschrift für Sozialreform*, vol. 29, pp. 297–337.

Teppe, K. (1977), 'Zur Sozialpolitik des Dritten Reiches am Beispiel der Sozialversicherung', *Archiv für Sozialgeschichte*, vol. 17, pp. 195–250.

Übelhack, B. (2002), *Ärztliche Ethik – Eine Frage der Ehre? Die Prozesse und Urteile der ärztlichen Ehrengerichtshöfe in Preußen und Sachsen 1918–1933*, Peter Lang, Frankfurt/M., in press.

Ullrich, C.G. (1995), 'Moral Hazard und gesetzliche Krankenversicherung. Möglichkeiten zur Mehrentnahme an Gesundheitsleistungen in der Wahrnehmung und Bewertung durch gesetzlich Versicherte', *Kölner Zeitschrift für Soziologie und Sozialpsychologie*, vol. 47, pp. 681–705.

Ullrich, C.G. (1996), 'Solidarität und Sicherheit. Zur sozialen Akzeptanz der Gesetzlichen Krankenversicherung', *Zeitschrift für Soziologie*, vol. 25, pp. 171–89.

Webber, D. (1988), 'Krankheit, Geld und Politik: Zur Geschichte der Gesundheitsreform in Deutschland', *Leviathan. Zeitschrift für Sozialwissenschaft*, vol. 16, pp. 156–203.

Webber, D. (1989), 'Zur Geschichte der Gesundheitsreform in Deutschland. II. Teil: Norbert Blüms Gesundheitsreform und die Lobby', *Leviathan. Zeitschrift für Sozialwissenschaft*, vol. 17, pp. 262–300.

Webber, D. (1992), 'Die kassenärztliche Vereinigung zwischen Mitgliederinteressen und Gemeinwohl', in Mayntz, R. (ed.), *Verbände zwischen Mitgliederinteresse und Gemeinwohl*, Bertelsmann Stiftung, Gütersloh, pp. 211–72.

Webster, C. (1988), *The Health Services Since the War*, vol. 1: *Problems of Health Care, the National Health Service Before 1957*, HMSO, London.

Webster, C. (1996), *The Health Services Since the War*, vol. 2: *Government and Health Care, the National Health Service 1958–1979*, HMSO, London.

Weindling, P. (1989), *Health, Race and German Politics Between National Unification and Nazism, 1870–1945*, Cambridge University Press, Cambridge.

Wiesenthal, H. (1981), 'Die Konzertierte Aktion im Gesundheitswesen: Ein korporatistisches Verhandlungssystem der Sozialpolitik', in Alemann, U. von (ed.), *Neokorporatismus*, Campus-Verlag, Frankfurt/M., pp. 180–206.

Wimmer, W. (1994), *'Wir haben fast immer was Neues'. Gesundheitswesen und Innovationen der Pharma-Industrie in Deutschland, 1880–1935* (Schriften zur Wirtschafts- und Sozialgeschichte, vol. 43), Duncker and Humblot, Berlin.

Wolff, E. (1997), 'Mehr als nur materielle Interessen. Die organisierte Ärzteschaft im Ersten Weltkrieg und in der Weimarer Republik 1914–1933', in Jütte (1997), pp. 97–142.

Zitzelberger, G. (1998), 'Gesundheit zum halben Preis', *Süddeutsche Zeitung*, no. 149, 2 July 1998, p. 21.

Chapter 4

Problems of Consent to Surgical Procedures and Autopsies in Twentieth Century Germany*

Cay-Rüdiger Prüll and Marianne Sinn

Introduction: The Problem of 'Informed Consent'

The present concept of 'informed consent' means that the patient has agreed to a specific diagnostic or therapeutic procedure after having been adequately informed about its risks and implications. This was by no means the situation at the beginning of modern medicine. Many studies on the history of medical ethics assure us that there was neither 'informed consent' in the present sense nor a discourse about this concept among physicians before 1947 (Winau, 1996: 13). In fact the *Nuremberg Code* of 1947 called for informed consent of all those participating in human experiments. However, 'information' and 'consent' with respect to human experimentation had already surfaced as an ethical issue during the late nineteenth and early twentieth centuries (Elkeles, 1996; Maio, 1996; Sauerteig, 2000; Weindling 2001). Although it is wise to avoid 'presentism' one may likewise seek the roots of the concept of 'informed consent' in the everyday clinical practice of that period.

The concept refers to a fundamental problem of the doctor-patient relationship. Before treatment starts, a decision-making process based on a discussion between the doctor and the patient or an exchange of non-verbal signals between the two parties takes place. There are as yet no satisfactory insights into how doctors and patients negotiate a mutual understanding of medical treatment in the context of scientific medicine. Studies on this subject have only recently been initiated (Maehle, 2000) or are currently in progress (Sinn).

Because of the many situations in the nineteenth and twentieth centuries where consent was important we compare two fields of medical activity:

* We wish to thank Andreas-Holger Maehle for important help and advice and also Ulrich Tröhler and all participants of the project 'Codification of Medical Ethics in England and Germany 1850–1933' for fruitful discussions.

pathology and surgery.[1] Pathology, or 'morbid anatomy' as it was also called in the nineteenth century, dealt with the organic basis of disease. In Germany it was closely linked to conducting post-mortem examinations. Clinical findings were compared to those present in the corpse. Apart from its routine work, pathology often focused on basic medical research. Surgery, by contrast, was much more concerned with daily therapeutic activity and the individual patient. Surgery and pathology allow us to study two different aspects of modern medicine. We will focus on Germany in the period between 1900 and 1933, i.e. until the end of the Weimar Republic and the start of the National Socialist dictatorship. As we shall see, lay criticism of professionalised medicine was increasingly articulated during this period. 'Information' and 'consent' became problematical. How did the disciplines of pathology and surgery deal with this challenge in the context of German medicine's hierarchical and highly paternalistic organisation, which was characterised by the dominance of powerful professorial heads of department, the so-called *Ordinarien*? This raises the question of the influence of institutionalisation on the issue of 'information' and 'consent'. Conflicts arose mainly from post-mortem examinations (autopsies) and, in surgery, from operations without consent.

In the following section we comment on the historical background of the problem as the disciplines of pathology and surgery developed. Secondly, we describe patients' attitudes and responses to the pathologist's and surgeon's actions. Thirdly, we analyse the reaction of pathologists and surgeons to patients' reluctance to adopt their views. As a fourth point, we focus on the legal aspects, which at times gained prominence because it was thought that the law would resolve disputes. Finally, we will summarise and analyse our conclusions.

The Historical Development of Surgery and Pathology

At the beginning of the twentieth century, pathology and surgery were each developing differently. German pathology was leading the field internationally in the nineteenth century. By 1850 pathological anatomy had become an integral part of medicine. The Berlin pathologist Rudolf Virchow (1821–1902) argued that 'morbid anatomy' was one of the most important cornerstones of scientific medicine. Virchow ascribed a threefold function to autopsy: as an accurate means to detect the cause of death; as a valid area of medical education; and as a basis for scientific research. Morbid anatomy was based on Virchow's theory of 'cellular pathology': pathological morphological changes were attributed to processes within the cell. Disease was seen as a process determined by the same laws that applied to physiological changes. Disease was not an independent entity. Instead, it was looked upon as the result of altered morphological structures. The microscope greatly enhanced both organ and tissue pathology. Post-mortem examinations became increasingly important for medical practice. However, because of its specialised scientific mission, pathology became detached from clinical practice. By 1900 nearly every German university had its own chair and institute for pathology (Prüll, 2002; Maulitz, 1993; Ackerknecht, 1953; Pantel and Bauer, 1990).

In the nineteenth century pathology helped to fill some blanks in the map of the human body. The morphological method made it possible to obtain information about structural changes in organs and their morbid appearances. However, after the turn of the century German pathology's significance for scientific medicine waned (Rössle, 1930). Clinical medicine increasingly gained independence from post-mortem examinations because of the invention of image-producing diagnostic technologies such as X-rays (1896). Moreover, the rapidly developing areas of bacteriology, pathological chemistry, biochemistry, and pathological physiology emphasised functional aspects and decisively changed the medical perception of the human body. Studies of dynamic processes within different parts of the body became much more important than the search for altered morphological structures. Medical research now focused on immunology, metabolic processes, and nutrition. Working with laboratory tests fostered the routine examination of small amounts of body fluids and pieces of tissue extracted from the living patient. Morbid anatomy had failed to give an answer to the ultimate question of the causality of disease. By now disease states could be measured and quantitatively differentiated. Although German pathologists acknowledged these new developments, they insisted that the morphological idea and their work in the post-mortem room were still at the very centre of pathology. Between 1900 and 1933 they were clearly on the defensive (Prüll, 2002).

By contrast, at the end of the nineteenth century, German surgeons had not yet maximised their popular success. Surgery, like other medical disciplines (Eulner, 1970), underwent a process of professionalisation. The field had been dominated by organ-centred thinking. It had been supported by anaesthesia from the mid-1840s, by asepsis, which had developed since the 1860s, and by pathological anatomy. Thus the surgeon's main task was the resection or removal of defective parts of the human body. Many new operations were introduced from the 1880s, for example the resection of the cancerous stomach by the Vienna surgeon Theodor Billroth (1829–1894) in 1881. However, only very few internationally standardised techniques existed. The success of operations depended mainly on the abilities of the individual surgeon. However, since the turn of the century, especially after the end of the First World War, the scientific reputation of surgery was strengthened. By incorporating the new physiological knowledge, surgery, with the help of laboratory medicine, focused on organ functions (Tröhler, 1993). New areas developed, for example the surgery of the thorax. This new thinking within surgery played an important role in more accurately determining the need for an operation and in controlling the intra- and postoperative phase. It may be said that the same functional ideas that put pathology into a defensive position, helped surgeons to develop more confidence in their abilities in their chosen specialties (Göckenjan, 1989).

Resistance against Autopsies and Surgical Interventions without Consent

In Germany the period between 1900 and 1933 was characterised by the resistance of patients against certain medical procedures and an increasing demand by patients for a role in the physician's decisions. The rapid process of industrialisation in the German Empire from about 1880 led to rising criticism of modern civilisation with its transformation of society and economics. One aspect of modern technical development was the rise of scientific medicine, which was perceived as 'materialistic' and impersonal. This criticism was fuelled by the German defeat of 1918. People increasingly favoured natural healers and alternative concepts in medicine. Between 1925 and 1935 scientific medicine in Germany seemed to be in a 'crisis' (Klasen, 1984).

By 1900 resistance to autopsy could look back on a long tradition. It had developed into a serious problem for German pathology since the 1850s when routine post-mortem examination was introduced. This 'resistance' was not politically organised but consisted chiefly of individual decisions. In 1886 Rudolf Virchow wrote: 'Every year the number of corpses on which relatives refuse to have an autopsy done increases.'[2] Resistance against autopsy was a cultural phenomenon rooted in religious as well as folk beliefs in life after death.[3] The latter, in particular, encompassed many different views and behavioural patterns regarding death and dying (Richardson, 1988; Löffler, 1975; Schepper-Lambers, 1992). Folk beliefs for example did not make a clear distinction between life and death. A corpse remained to some extent 'active' and powerful. It was therefore necessary to obey certain rules and to perform certain rituals, which had mainly three purposes. Firstly, they 'cared' for the deceased whose path from earth to heaven had to be protected and who had to be relieved of all burdens. Secondly, they were important for the survivors who had to protect themselves against being disturbed by the deceased if anything had gone wrong (e.g. in the distribution of the deceased's possessions on earth or in the burial rites). Thirdly, death and dying were events with a potential for good or evil magic. The corpse could be a source of magic power. Rules and rituals preserving the integrity of the human corpse were therefore deemed essential. Hence, to perform a post-mortem was to 'desecrate' the corpse (Prüll, 1996: 175–6).[4] Several sources show that such beliefs had a strong impact on people in the first half of the twentieth century and still do today.[5]

Resistance against autopsies became stronger from the 1880s and reached a peak in the 1930s. The most significant problem was that of 'information' and 'consent'. This is illustrated by the following cases. In 1931 a woman complained about the autopsy that had been performed on her husband in the Charité-Hospital in Berlin. In the mortuary she had lifted his shroud and discovered incisions. She subsequently wrote to the Ministry of Science complaining that the body had been cut up in a criss-cross fashion and that all the organs had been removed. She also claimed that her husband had been 'slaughtered alive' for the sake of scientific knowledge. The woman informed the 'League of Human Rights'.[6] Her complaints seem to have been highly exaggerated, and they were dismissed with the comment that the woman was mentally ill.[7] However, the correspondence with the Ministry and the Charité sheds light on the attitudes of the physicians and pathologists involved. The physicians claimed there had been no objection to the post-mortem from the relatives and that it had been ordered purely out of scientific interest.[8] Talking to the woman, the pathologist defended his work with the remark: 'Do what you like if you wish to take action against us. I have to protect my colleagues. Good afternoon!'[9] The woman noticed that the pathologist was very nervous, much more than she was herself.[10]

A second example from the Berlin Charité-Hospital provides a far more detailed picture of the difficulties encountered with autopsies. In November 1929 a well-known French movie star and world-champion wrestler of African origin died of leukaemia in the hospital. Although his wife had specifically prohibited a post-mortem examination, the pathologist opened the corpse. Only by chance did the widow discover this blatant disregard of her husband's wishes. Dressed in black she entered the dissection room of the Pathological Institute, ignoring any resistance she encountered, and caught the pathologist at work.[11]

The woman complained about the doctors, especially about the pathologist. Her husband, she said, had been a member of the Roman Catholic Church and had dreaded an autopsy. The newspaper *Neue Berliner Zeitung* reported the case and published the widow's accusation. The headline about 'strange corpse rituals' in the Charité-Hospital reflected common fears. The wife of the wrestler claimed compensation for the unlawful mistreatment of her husband, and the case was brought to the attention of the Prussian Ministry of Science. The Director of the Pathological Institute, Robert Rössle (1876–1956), defended himself. He said that he had told his assistant to perform a post-mortem 'because clinically it was an especially important case and, because there was a written statement of clinical interest, the objection was invalid'. Rössle based his defence on section 9 of the Charité-Hospital's regulations on the handling of corpses, which said that any corpse could be opened if this was of scientific interest. As a consequence, the Ministry dismissed the protest of the wrestler's wife.[12] This case illustrates that pathologists were clearly more interested in using the human body for medical science than in acknowledging lay views on respecting the dead.

From the end of the nineteenth century patients also showed an increasingly critical attitude towards surgical operations. As with pathology, 'information' and 'consent' played a prominent role. Again we would like to illustrate this with specific cases. In 1894 the father of a seven-year-old girl refused a partial amputation of the forefoot of his daughter who suffered from tuberculosis of the bone. Despite the father's protests the surgeon performed the operation and the girl recovered.[13] This case is an example of a patient's general demand for informed consent. Although the operation was successful the patient's father protested as a matter of principle, because the surgeon had not sought his consent.

A second type of a patient's request for informed consent can be illustrated by the following case. In 1912 a patient refused to pay the surgeon because an ear operation had led to paralysis of an auditory nerve and the facial nerve. The patient argued that the surgeon should have informed him about the risks of the operation.[14] In both cases, the patient or relative did not give consent. But in the second case the patient also complained about the lack of sufficient information.

There is a third type of patient protest. In 1932 a 67-year-old man gave his consent to an operation on his hands. The operation failed and led to amputation of one finger. The man did not complain because of lack of consent or information but because he felt he had been entirely misinformed. The patient stated that the surgeons had promised him complete recovery.[15] These cases show that conflicts between patients and surgeons involved lack of consent, lack of information, and incorrect information. The outcome of an operation was an important factor in such conflicts.

Patient protests in the fields of pathology and surgery differed in their significance. The pathologists' work, which aimed at obtaining basic knowledge about the human body, essentially meant doing scientific research. Because of this, and because autopsies clashed with cultural beliefs and practices, resistance against post-mortem examinations ultimately meant questioning scientific medicine and its principles. Surgeons, on the other hand, were confronted with objections to specific operative procedures and with protests about misinformation or lack of information. Patients' criticism tended to focus on the result of a particular treatment.

Reactions of Pathologists and Surgeons to Patients' Demands for 'Information' and 'Consent'

As pathologists in Germany started to perform autopsies on a routine basis they had to overcome people's resistance against this practice. They began formulating various arguments to justify the need for post-mortems. These arguments were repeated over and over, becoming a kind of standard verbal response to any criticism from outside the scientific community. The defence relied mainly on Virchow's notion of the threefold function of morbid anatomy. Remarkably, Virchow himself never publicly responded to any objections made by someone's next of kin concerning autopsy. For him the purpose of post-mortem examinations was self-evident.

Especially from 1900 onwards pathologists relied on Virchow's arguments in justifying the significance of autopsies and in answering criticism. It was important to obtain the relatives' consent to an autopsy, but it was just as important to provide the information necessary for obtaining such consent. Even in the First World War, when post-mortem examinations could be conducted on soldiers without the consent of relatives, there was resistance to autopsies, and pathologists tried to persuade people to accept them as a scientific need. The army pathologist Hugo Häßner wrote in 1916 that offering one's body for autopsy was 'definitely the pinnacle in comradeship and Christian love for one's neighbour even in death!' (Häßner, 1916: 309–10). Pathologists attempted to familiarise soldiers and other military personnel with autopsies by having them present when they opened the corpses (Häßner, 1916: 310; Ricker, 1921: 334; Prüll, 1996). They hoped that people who had been educated in this way would not have any problems with autopsies in the future. The significance of giving information can also be illustrated by an initiative of the Freiburg pathologist Ludwig Aschoff (1866–1942), who was the leading figure in German pathology in the first half of the twentieth century.[16] Although Aschoff, like his colleagues, was a strong advocate of autopsies he thought it important to keep in touch with the public's view on this matter. In October 1925 he gave a public lecture about the importance of post-mortem examinations and animal experimentation for public health and social welfare. Aschoff admitted that autopsies belonged to the category of things 'which are actually at odds with our natural human feeling' (Aschoff, 1925: 151),[17] thus showing that he took feelings into account. He accepted the unease connected with post-mortems, feelings which he had experienced himself and which were also expressed by medical students. Since objections against autopsies were fundamental to Aschoff, he tried to find a form of co-existence between popular beliefs and scientific medicine. Aschoff also hoped for a serious discussion: 'It is obvious that one can speak with total frankness only if the listener has the necessary gravity and the will to understand'.[18] He provided a general description of autopsies and gave also more specific information about its means by carefully describing the technique of opening the body and by pointing out the differences between autopsy and dissection: 'The corpse is not cut up in criss-cross fashion; instead, only one incision is necessary.'[19] Aschoff also emphasised the (relative) intactness of the body, which should be preserved whenever possible. His efforts to obtain a consensus between pathologists and relatives were not typical of German pathology, but Aschoff was a key figure in the field in his time. Moreover, his comments do at least show the basic approach used, which was to attempt to obtain the relatives' consent through giving relevant information.

Like the pathologists, surgeons wanted to avoid any problems with obtaining consent. The Swiss Theodor Kocher (1841–1917) argued in 1907 that surgeons should be free to choose where and how to operate.[20] Three years later his German colleague, Albert Krecke, noted that information about the necessity and harmlessness of surgical procedures convinced the less intelligent patient. The surgeon should therefore simply say that there was a tumour in the breast, in the uterus, or in the stomach, which could 'easily' turn malignant. There would then be no problem with consent.[21] In 1923 Krecke stated that patients had the right to be informed about their illness, but only to the point where it would not disturb their mental equilibrium.[22] In the 1920s surgeons tended to keep a diagnosis of cancer secret in order not to worry the patient. Contrary to the custom of Anglo-American surgeons, this was considered a humane tradition on the Continent.[23]

On the other hand, in 1933, surgeons proposed showing patients pictures of the early and late stages of cancer to illustrate what would happen without early surgical intervention.[24] In 1930 the surgeon Hugo Stettiner advised that the problem of the patient's psyche could be circumvented with certain drugs and by deceiving the patient about the exact day and hour of an operation.[25] Even the surgeon Erwin Liek (1878–1935), one of the critics of modern scientific medicine and a supporter of natural healing methods,[26] summarised the view common among his colleagues when he stated, in 1926, that every patient should be 'guided' by the surgeon.[27] This idea was based on the belief that patients were unable to understand the significance and implications of a surgical intervention. Obtaining the patient's fully informed consent was deemed impossible. Instead, compliance was sought through trust in the surgeon's best intentions. By the end of the 1920s very few voices in German surgery emphasised the need to 'inform' patients about details of their disease and therapy. However, the surgeon Fritz König observed in 1931 that more and more patients were asking for information. He claimed that it was the surgeon's duty to provide this information but to reduce it to the most basic, necessary facts.[28]

As lay criticism of autopsy and surgical procedures had different causes and patterns, pathologists and surgeons tackled the problem of informed consent quite differently. Pathologists discussed the issue of consent and provided general information in the sense of educating the public. But they did not want to acquiesce in co-operating with the relatives. They merely wished to overcome their resistance. Pathologists responded fiercely to relatives' protests against post-mortem examinations, because such protest implied a general criticism of their scientific method. They were especially alarmed because pathological anatomy was beginning to lose significance for medical progress.

For surgeons the need to inform patients and to obtain their consent could become a disruptive factor in daily practice. Therefore, the demands of patients had to be pushed away as quickly as possible. Our analysis of monographs on general surgical problems and of articles in surgical journals shows that 'information' and 'consent' were of no major concern to surgeons. They felt quite self-confident about their therapeutic achievements and saw surgery as an increasingly important discipline. Surgeons did not take patients' views very seriously. In a paternalistic manner they thought themselves in a superior position. The patient was believed to be unable to understand the surgeon's language and was not treated as a partner with equal rights. Informing patients about an operation was of very minor importance to the surgeon as long as they complied.

Pathology, Surgery, and the Law: Consequences of the Conflict

In Germany autopsy was not covered by any specific law. 'Information' and 'consent' depended on the attitude of the individual pathologist. During the nineteenth and twentieth centuries the decision on how to deal with the corpse of someone who had died in hospital was based on the regulations of the relevant institution. These regulations included specific instructions for pathologists. However, they were not standardised and differed from one hospital to another, depending on the governmental guidelines in the relevant German state (Oberhoff, 1935: 37). Moreover, the regulations changed with new social and political conditions. The Berlin Charité-Hospital provides an example.

In 1856 Rudolf Virchow succeeded in establishing a regulation that permitted post-mortems if the relatives did not object within 18 hours. In 1906 this period was reduced to 12 hours (Doerr, 1979: viii; Orth, 1906: 824). The regulation was again changed during the Revolution of 1918–1919. In December 1918 the democratic Ministry of Science instructed the Charité administration to comply with every objection against an autopsy raised by the relatives, if the administration, or the director of the Pathological Institute, did not succeed in persuading them to withdraw their objections (Lubarsch, 1931: 340). The pathologist Otto Lubarsch (1860–1933), director of the Berlin Pathological Institute and a successor in the Virchow chair, saw this instruction as an attack against his authority. Lubarsch fought the new regulation, arguing that education and research in pathology would be severely damaged. He convinced his colleagues in the Medical Faculty to sign a petition against the instruction. However, the new regulation remained in force, and he received no reply from the Ministry of Science.

Lubarsch interpreted the Ministry's silence as approval. Without any consent he performed autopsies on a number of persons who had been killed in Berlin street fights in January 1919. The Ministry now reminded him to comply with the new regulation. However, he objected to this, referring to his duties regarding medical education and research. He simply refused to accept the Ministry's admonition. Again he received no further reply from the Ministry. Lubarsch then decided to proceed with post-mortems without waiting for the consent of relatives. In the following years he was reminded several times of the Ministry's guidelines, but he referred then to his former arguments. Finally, in Lubarsch's last year as director, the regulation was changed (Lubarsch, 1931: 340–41). In 1929 his successor Robert Rössle could refer to section 9 of the Charité regulations which restricted the right to object to a post-mortem on a deceased relative. Berlin and the Charité were typical examples of the situation in Germany: there was no defined legal framework for the management of autopsies; there were only local hospital regulations that could be changed by the governments of the German states. However, pathologists pushed for a solution to end discussions (Prüll, 2002).

The lack of a law on autopsies became a problem for the courts. Lawsuits involving post-mortems were possible only under the terms of the legal systems of the German states. After 1871 the civil and penal codes of the German Empire provided no more than a basic framework for the regulation of autopsies. A decisive factor was the German Supreme Court (*Reichsgericht*) which set precedence for case law. Between 1880 and 1945 it arbitrated on seven cases dealing with the handling of corpses.[29] One criminal case from September 1930 illustrates the dominant position of the Supreme Court. The deputy director of the University Clinics of Königsberg performed a post-mortem on a woman without obtaining the permission of her next of kin. The physician defended himself, emphasising that he had not been aware of any objections by the relatives. The Supreme Court discussed the case at remarkable length. Arguments focused on the question whether a corpse was legally an 'object' (article of property) and whether the deputy director of the Clinics had in this sense damaged the property of others. Based on court decisions in Germany since 1903 the Supreme Court rejected this view, partly because the personal rights of the dead were recognised in law. It is interesting that the Supreme Court did not wish to decide on principle whether a corpse was an article of property or not. The fact that nobody really was in possession of a corpse and that it thus could not be defined as the property of another person was of decisive importance in this case. The court decided that the physician had not damaged an 'object' that was in the possession of another person, and he was therefore acquitted.[30]

This case is remarkable because the Supreme Court could not clarify the legal situation: 'A specific legal regulation concerning the question discussed here is, of course, lacking. But because of common attitudes and customs a legal position has habitually developed which treats the human corpse under exceptional legal premises because of its peculiarities. In any case, no right of property concerning a corpse comes into existence by death alone.'[31] Nevertheless, a customary right of the next of kin had developed. The next of kin decided where and how the funeral was performed and, importantly, whether to reject any intervention on the corpse. In this way, the Supreme Court confirmed a view that had previously been discussed by legal experts: the corpse may be in the care of a person or an institution, for example the hospital, but the relatives have the 'preferential right of appropriation' (*vorzugsweises Aneignungsrecht*) over the corpse. However, because the corpse was no one's property, those who had custody of it could not be accused of (useful) procedures carried out on the corpse, at least not if they had not seized control of the corpse by 'unauthorised encroachment' (*unbefugter Übergriff*). Because the corpse was not someone's property, the pathologist could not be accused of damaging property while performing an autopsy, and because the corpse was no longer a living being, a post-mortem could not be deemed battery (*Körperverletzung*) either. Because the pathologist did a post-mortem for scientific reasons, opening a corpse did not constitute 'gross mischief' (*grober Unfug*). Therefore, a physician who performed an autopsy could not be punished according to penal law.[32] Yet even in 1928 pathologists tended to acknowledge the wishes of the deceased or relatives (Seesemann, 1928). Although some pathologists called for a law permitting post-mortems on a broad scale and although lawyers for the most part supported the pathologists' aims, there were no amendments to the law in this respect nor any specific autopsy legislation during the Nazi period (Brugger and Kühn, 1979: 124). The situation remained open to the views and wishes of the pathologist and the relatives, depending on the case and the special circumstances. Autopsy was a sensitive issue because changing its legal status would have meant deciding in favour of one of the opposing systems of belief: that of scientific medicine or that of folk customs and religious belief. To this day German pathologists are trying to avoid the requirement of 'information' and 'consent' by demanding autopsy legislation.

Regarding surgery, likewise, no laws on patient 'information' or 'consent' pertained. Conflicts were avoided because nineteenth century hospital regulations clearly subjected the patients' will to the plan of action of the physician or surgeon. A debate did not start until 1892 when legal experts in the German-speaking countries discussed whether performing surgical procedures without the patient's consent constituted assault and battery. Andreas-Holger Maehle has shown how these discussions led to the acceptance of the 'battery theory' of surgical interventions during the first two decades of the twentieth century. Although on the face of it this strengthened the position of the patient, the paternalistic concept of the patient's 'implied consent' (*stillschweigende Zustimmung*) hampered the efforts of those who wished to give more weight to the patient's wishes (Maehle, 2000).

The German Medical Courts of Honour dealt mostly with intra-professional problems (Maehle, 1999: 321–9; Übelhack, 2002), but the above mentioned conditions can be identified in lawsuits against surgeons. In 1894, following a lawsuit, the Supreme Court considered surgical intervention without consent to be battery.[33] This view caused considerable controversy among lawyers. Some argued that the law could not on the one hand permit operations by giving doctors the licence to heal and on the other hand treat surgical procedures as battery. Others argued that the patient's consent had to be obtained before an action which constituted battery could turn into a legal therapeutic procedure.[34] In the following period the Supreme Court – as in the case of autopsy – focused on the issue of consent. Its decisions in this respect favoured the surgeons. In 1907, when dealing with a case of an unsuccessful operation on an under-age patient, the Court applied the concept of so-called 'implicit consent'.[35] This meant that if the patient did not explicitly oppose the treatment suggested by the surgeon, the latter had the right to carry it out. This 'implicit consent' was seen as legally equivalent to so-called 'explicit consent' (*ausdrückliche Zustimmung*). If there had been no consent at all, surgeons had to face a charge of battery.

In March 1912 some new aspects were considered. A patient was operated on because of hearing problems. The surgical intervention led to deafness and the patient refused to pay the fee. A County Court found the surgeon guilty because he failed to inform the patient of the risks of the operation. However, the Supreme Court overturned this judgement with the argument that the surgeon had not been obliged to inform the patient. The Supreme Court justified its decision by arguing that too much information might deter patients from necessary operations or that additional worries might hamper recovery.[36] Thus, in 1912 'informing' the patient was mentioned, but there was no change in the legal status established in 1894. After 1912 there were several more cases dealing with lack of consent to surgical procedures. In May 1931, for the first time the Supreme Court mentioned a doctor's duty to inform the patient about an operation.[37] From now on one could speak of 'informed consent', but in reality surgeons avoided 'informing' their patients. They only focused on their 'consent'. After 1933 the development of the law towards 'informed consent' proceeded under restrictions. Only in 1957 did the Supreme Court of the Federal Republic of Germany elaborate the meaning and use of 'informed consent' as we understand it today (Roßner, 1998: 46–53).

It is remarkable that between 1900 and 1933 – contrary to the discussion of human experimentation (Winau, 1996: 25) – the legal situation regarding operations was not much discussed among surgeons. Only occasional remarks on the topic can be found. The comment of the Munich professor of surgery, Ottmar von Angerer (1850–1918), on the 'battery discussion', that 'a doctor who skilfully removes with one cut a malignant tumour' should not 'be treated on the same level as a rowdy' (1899)[38] seems to have been rather exceptional.

This is surprising as there had been increasing interest in legal questions among German physicians since the turn of the century. Topics such as medical treatment and bodily harm, and the role of information and consent of the patient, were increasingly discussed in general medical journals such as the *Deutsche Medizinische Wochenschrift* and *Ärztliches Vereinsblatt*.

There were no such discussions in surgical journals. One has to conclude that surgeons showed little interest in the legal problems relating to 'information' and 'consent'. Only in 1929 the newly founded journal 'The Surgeon' (*Der Chirurg*) introduced a special column for legal questions.[39] In 1937 the German Surgical Association (*Deutsche Gesellschaft für Chirurgie*) for the first time discussed the legal foundations of surgical work. At the relevant meeting Nicolai Guleke (1878–1958), professor of surgery at the University of Jena, gave a lecture on the consent of the patient and on giving information before surgical procedures.[40] He claimed that the surgeons themselves should decide whether and how to inform the patient because only they were able to judge the patient's capacity to understand any explanations. Guleke emphasised that in his view there was no such capacity at all on the patient's side.[41]

It is difficult to detect the reasons for surgeons' increasing interest in consent and information from the mid 1930s. One reason may have been the growing knowledge in physiology, microbiology, and pharmacology that now questioned the superiority of surgery. Surgeons may have experienced the same problems that the pathologists encountered 40 years earlier.

Pathology and surgery as medical disciplines developed differently. Patients' claims on 'information' and 'consent' were also different in the two disciplines. Pathologists and surgeons maintained different strategies for solving the question of consent and/or information in regard to the courts. Pathologists were faced with the lack of a law on autopsies and were calling for legislation that would enable them to perform post-mortems without restrictions. This situation has not changed up to the present day.

Surgeons, until 1933, did not feel under pressure to extensively discuss the legal issue of consent or information of their patients. Obviously they did not consider the existing legal situation appropriate. Lawsuits in connection with conflicts over the issue of consent were usually decided in favour of the surgeon, although surgeons had sometimes performed operations without the patient's consent and without previously informing the patient. Neither did surgeons care very much about the problem of 'medical intervention as battery' (*ärztliche Eingriffe als Körperverletzung*), although this problem caused constant annoyance in the medical profession. The profession was worried about a possible violation of doctors' honour, because the concept of surgery as battery put them legally in one group together with – as physicians said – the 'ordinary criminal' (*gemeiner Verbrecher*).[42]

Summary and Conclusion

Our comparative analysis of the attitudes of German pathologists and surgeons towards informed consent sheds light on the problem within scientific medicine more generally. Both disciplines, between 1900 and 1933, followed a paternalistic tradition. Medical professionals did not consider the patient to be a partner with whom to negotiate appropriate treatment. German pathology – in the context of its declining status in medicine – had to defend autopsy as a legitimate method against growing public resistance. Pathologists recognised the need for 'information', which they regarded as helpful in obtaining 'consent'. This seemed to be necessary because criticism of post-mortems meant criticism of the whole discipline and of scientific medicine itself. Pathologists wanted to halt these objections by getting rid of any obligation to negotiate with relatives under an autopsy law.

By contrast, surgery did not have to face such criticism. It was a discipline of increasing importance and self-confidence. Since surgeons thought they had the power and knowledge to analyse human diseases as well as to cure them, they did not see why, from their position of superiority, they should have to deal with the problem of seeking patients' consent. Information and consent-seeking was seen as something that could cause disruption in daily practice. Many surgeons simply ignored the issue. In contrast to the pathologists it was not really necessary for them to look for a legal solution for 'information' and 'consent'.

This means that the answer to the question whether and how scientific medicine tackled the problems of 'information' and 'consent' seems to depend on the specialism itself. Our findings show that the history of specialisms and their institutions, and the shaping of the mental attitudes of its representatives as a social group, strongly influenced the implementation of 'information' and 'consent' in early twentieth century medicine. Patients criticised basic preconditions of a certain medical discipline, such as the unimpeded use of the scalpel in surgery, or challenged preconditions of scientific medicine, for example autopsies in pathology. Resistance of members of medical disciplines to the recognition of patients' wishes depended on the discipline's power. Surgeons could ignore such wishes; pathologists could not. Both medical fields were able to defend their position. Although negotiations mainly dealt with the problem of 'consent', the question of 'information' played an increasing role since 1900.

Looking at 'information' and 'consent' in recent medical history, it appears difficult to recognise actual 'informed consent', as often there was a large gap between the expert's and the patient's knowledge and experience. Nevertheless there is a crossover. Questions concerning the patient's personal situation in life need to be discussed on this basis. Such questions cannot be answered by medical expert knowledge alone. Therefore discussion is not only possible but also necessary. If patients are seen as responsible and taken seriously, if their arguments are heard and their views explored by the doctor or scientist, a fruitful, informed consent should be the result. Above all, the story of how German pathologists and surgeons responded to demands for 'information' and 'consent' shows that medicine is not only influenced by scientific but also by social developments.

Notes

1 The study of informed consent in German and British surgery is part of a project on the codification of medical ethics in Britain and Germany between 1850 and 1933, organised by Andreas-Holger Maehle, Department of Philosophy, University of Durham, United Kingdom, and Ulrich Tröhler, Institut für Geschichte der Medizin, University of Freiburg, Germany. Marianne Sinn is working on consent in German surgery between 1900 and 1933. In this essay her work is combined with results of Cay-Rüdiger Prüll's work on the history of pathology in Britain and Germany in the twentieth century.

2 'Mit jedem Jahre nimmt die Zahl der Leichen zu, welche durch die Angehörigen der Sektion entzogen werden'. Virchow, 1886, p. 295.

3 Also today there are approaches – even from the clerical side – that reduce the problem of resistance against autopsies to mere religious feelings. This argumentation is far too short. See Böckle, 1983, pp. 1–2.

4 On cultural and anthropological resistance against autopsy see the articles by P. Geiger, 1932–33 and 1936–37.

5 In historical perspective: Bächtold, 1917; Kronfeld, 1915; Bohne, 1932; Oberhoff, 1935, p. 5; Grynaeus, 1993; Probst, 1994. On contemporary views regarding autopsy see for example the comments within the trade of organs discussion in Germany in 1993 in *Der Spiegel*, no. 51 (1993), pp. 10, 12. The corpse is dealt with as 'living person' in Heym, 1984. Countless movies and pieces of literature even today deal with the resurrection of dead persons because of their mistreatment in life.

6 'Liga für Menschenrechte', see: Complaint T.K. to the Minister of Culture, Berlin, 20 July 1932 (3 pp.) in: Verfahren bei der Behandlung der in dem Charité-Krankenhause Verstorbenen, insbesondere Sektionen; Dezember 1929 bis März 1938, Bundesarchiv Berlin, Abteilung Lichterfelde (BArchiv Berlin), Reichsministerium für Wissenschaft, Erziehung und Volksbildung (REM), vol. 2, no. 2697 (no page numbers); A.K. to the Minister of Science, Berlin, no date (3 pp.), ibid.

7 See the relevant letters, ibid.

8 Prof. Kauffmann, Dr Stroebe, II. Medical Clinic of the Charité-Hospital to the Charité-Administration, Berlin, 7 December 1931 (2 pp.); The Ministry of Science to Mrs K., Berlin, 26 January 1932 (2 pp.), ibid.

9 'Unternehmen Sie gegen uns was Sie wollen, ich muß meine Kollegen schützen. Guten Tag', see: Mrs A.K. to the Ministry of Science, no date, pp. 3-4, ibid.

10 Ibid., pp. 3-4.

11 The Ministry of Science, Notice, 27 November 1929, ibid.

12 '... da es sich um einen klinisch besonders wichtigen Fall handelt und der Einspruch bei dem schriftlich geäußerten klinischen Interesse dadurch hinfällig geworden war', see: Robert Rössle to the Charité-Administration, Berlin, 27 November 1929, ibid. See also ibid.: The Ministry of Science, Notice, Berlin, 27 November 1929; Max Lenz, 'Sonderbarer Leichenkult in der Charité. Obduktion trotz Einspruchs der Hinterbliebenen. Klinisches

Interesse geht vor', *Neue Berliner Zeitung*, vol. 11, no. 274, 22 November 1929; Charité-Administration. Ordnung über das Verfahren bei der Behandlung der in dem Charitékrankenhause Verstorbenen, Berlin, 10 January 1928.

13 *Entscheidungen des Reichsgerichts in Strafsachen*, vol. 25 (1894), pp. 375–89. See also chapter 2 by Maehle, this volume.

14 *Entscheidungen des Reichsgerichts in Zivilsachen*, vol. 78 (1912), pp. 432–6.

15 *Urteil des Kammergerichts Berlin vom 24.11.1932*, in Richard Goldhahn and Werner Hartmann, *Chirurgie und Recht*, Enke, Stuttgart, 1937, p. 42.

16 For Aschoff's life see: Aschoff, 1966; Büchner, 1943; Büchner 1957; Seidler, 1976; Seidler, 1991, pp. 207–10, 270–72, 333–4; Buscher, 1980; Fischer, 1964; Meeßen, 1975, pp. 19-31; Prüll, 1995; Benaroyo, 1998.

17 'Dinge, die unserem natürlichen menschlichen Empfinden eigentlich entgegenstehen, mit denen wir uns nicht ohne weiteres abfinden können', Aschoff, 1925, p. 151.

18 'Es ist ganz selbstverständlich, daß man mit voller Offenheit nur dann sprechen kann, wenn man bei der Zuhörerschaft den nötigen Ernst und auch den Willen zum Verstehen findet', ibid., p. 153.

19 'Die Leiche wird keineswegs kreuz und quer zerschnitten, sondern es ist nur ein einziger Schnitt, der notwendig ist', ibid., p. 158.

20 'Man soll dem Chirurgen völlig freie Wahl für den Ort und die Art der Ausführung der Operation lassen.' Kocher, 1907, pp. iv–v.

21 '... bietet es im allgemeinen keine Schwierigkeiten auch weniger intelligente Kranke zu einem chirurgischen Eingriff zu bestimmen. Man braucht den Kranken nur zu sagen, dass es sich um eine Geschwulstbildung in der Brust, im Uterus, im Magen handle, die später leicht bösartig werden könne, und man wird mit der Zustimmung zur Operation keine Schwierigkeiten haben'. Krecke, 1913, pp. 714–15.

22 'Und muß der Kranke über jede Störung in seinem Organismus genau unterrichtet sein? Gewiß hat er ein volles Recht darauf, zu erfahren, was ihm fehlt, aber nur unter der Einschränkung, daß sein seelisches Gleichgewicht nicht gestört wird.' Krecke, 1926, pp. 6–7.

23 'Noch eine Bemerkung allgemeiner Natur, die mit dem engeren Thema der Arbeit zusammenhängt: Die Schule Hocheneggs wahrt absolut den auf dem europäischen Kontinent üblichen Brauch, dem krebskranken Patienten seine Diagnose mit allen Mitteln zu verheimlichen. Bekanntlich steht dieses Vorgehen im Gegensatz zu jenem englischer und amerikanischer Ärzte. Dem letzteren wird von manchen Seiten nachgerühmt, daß es bei gleichzeitiger Betonung der Heilbarkeit oft die Möglichkeit gibt, Patienten noch einer Radikaloperation zuzuführen, die sich einer solchen sonst entzogen hätten.' Fuchs and Panek, 1928, p. 263.

24 'Wenn schon eine Propaganda betrieben werden soll, so erscheint es logischer, Abbildungen von Früh- und Spätstadien zu veröffentlichen und dazuzuschreiben: Wenn du dich nicht in dem ersten Stadium behandeln oder operieren läßt, dann wird die Erkrankung so, wie sie dir das zweite Bild zu erkennen gibt.' Hahn, 'Können wir durch organisatorische Maßnahmen in der Krebsbekämpfung einen wesentlichen Einfluß auf die Krebssterblichkeit erwarten?', *Archiv für klinische Chirurgie*, vol. 177 (1933), p. 45.

25 'Jedenfalls wird auch durch das Pernokton wie durch das Avertin die Psyche vollkommen ausgeschaltet, besonders wenn es gelingt, die Patienten über Tag und Zeitpunkt der Operation hinwegzutäuschen.' Hugo Stettiner, 'Chirurgie. Fortschritte der gesamten Medizin', *Deutsche Medizinische Wochenschrift*, vol. 56 (1930), p. 1137.

26 On Liek see Schmiedebach, 1989.

27 'Jeder Kranke, auch der Höchststehende will geleitet sein.' Liek, 1926, pp. 37–47.

28 'Selbst ein erfahrener Chirurg kann sich in der Abschätzung irren. Soll er nun aber all diese Überlegungen vor dem Patienten ausbreiten oder soll er das ganze Risiko auf sich nehmen

und dem Kranken zur Operation raten? Astley Cooper stand noch auf dem Standpunkt: "Es kann nie gut sein, die Kranken wissen zu lassen, was man zu ihrer Heilung beabsichtigt, die sehr oft dadurch vereitelt wird." Aber der kluge englische Chirurg lebte in einer anderen Zeit; noch war die chirurgische Behandlung weniger kompliziert; aber auch die Kranken dachten anders. Heute wollen sie meist sehr viel mehr von dem wissen, "was man zu ihrer Heilung beabsichtigt", und auch der Chirurg ist ihnen sowohl eine weitgehende Aufklärung schuldig – wie auch sich selbst, wenn schlechte Ausgänge möglich sind. Dabei darf er nie so weit gehen, daß er den Patienten vor der notwendigen Operation zurückschreckt. Oft wird er, indem er den Kranken bis zu einem notwendigen Grad, die Angehörigen aber weitestgehend aufklärt, das Richtige treffen, um auch sein Gewissen zu beruhigen.' König, 1931, p. 25.

29 'Genügt es, wenn die ärztlichen Sachverständigen, welche sich bei Abgabe ihres Gutachtens zugleich über ihre Wahrnehmungen betreffs des thatsächlichen Leichenbefundes, der vorgefundenen Verletzungen u.s.w. ausgelassen haben, nur als Sachverständige, nicht auch als Zeugen beeidigt werden?', *Entscheidungen des Reichsgerichts in Strafsachen*, vol. 2 (1880), pp. 389–90. '5. Enthält die in der Hauptverhandlung mündlich erfolgte Bestätigung eines auf den Sektionsbefund gegründetetn schriftlichen Gutachtens seitens der begutachtenden Ärzte zugleich die mündlich erfolgte Bestätigung des verlesenen Protokolls über die Leichenöffnung?'; '6. Darf das Protokoll über die Leichenöffnung, im Gegensatz zu demjenigen über die Leichenschau, in der Hauptverhandlung verlesen werden?'; '7. Ist der Grund der Verlesung im Sitzungsprotokolle anzugeben?', ibid., vol. 2 (1880), pp. 153–60. '1. Was ist unter "Beiseiteschaffen" eines Leichnams zu verstehen?'; '2. Wann ist ein Leichnam "ohne Vorwissen der Behörde" beiseite geschafft?', ibid., vol. 28 (1896), pp. 119-22. 'Kann in dem unbefugten Herausnehmen einer Leiche aus der noch offenen Gruft, in der sie beigesetzt war, ein Vergehen gegen § 168 St.G.B.'s gefunden werden ?', ibid., vol. 28 (1896), pp. 139–41. 'Der Veranstalter einer "politischen Demonstration" bei einem Leichenbegängnis als Veranstalter eines nicht gewöhnlichen Leichenbegängnisses', ibid., vol. 45 (1912), pp. 85–8; 'Vorraussetzungen für die Lesbarkeit eines Leichenschau-protokolls', ibid., vol. 53 (1920), pp. 348–9. 'Kann an einem menschlichen Leichnam durch unbefugte Leichenöffnung Sachbeschädigung begangen werden?', ibid., vol. 64 (1931), pp. 313–16.

30 Ibid., pp. 312–16.

31 'An einer ausdrücklichen gesetzlichen Regelung der zur Entscheidung stehenden Frage fehlt es freilich. Es hat sich aber gewohnheitsrechtlich auf Grund der Volkssitten und =Gebräuche eine Rechtslage herausgebildet, nach welcher der Leichnam des Menschen um seiner Besonderheiten willen eine eigenartige rechtliche Behandlung erfährt, jedenfalls aber an ihm nicht schon durch den Tod ein Eigentumsrecht zur Entstehung gelangt', ibid., p. 315.

32 Bauer, 1929, pp. 20–27. The decisive paragraphs in the German penal law (Strafgesetzbuch) concerning autopsies are § 168 (theft of a corpse), § 367 (removing parts of a corpse). Furthermore, the following paragraphs play a certain role: § 303 (damaging an object), § 223ff (battery), § 360, 11 (mischief). See Oberhoff, 1935, pp. 6–29; Seesemann, 1931; Bohne, 1932.

33 *Entscheidungen des Reichsgerichts in Strafsachen*, vol. 25 (1894), pp. 375–89.

34 On medico-legal debates at the end of the nineteenth century, see Eser, 1990; Laufs, 1986.

35 Decision of the German Supreme Court of June 21, 1907: 'Die stillschweigende Einwilligung entspricht der ausdrücklichen', in: Bestand des Bundesarchives R 3002 (Reichsgericht)/Aktenzeichen III 465/05.

36 *Entscheidungen des Reichsgerichts in Zivilsachen*, vol. 78 (1912), pp. 432–63.

37 Reichsgerichtsurteil vom 19.5.1931, *Deutsche Medizinische Wochenschrift*, vol. 57 (1931), p. 2113/Archiv III 202/30.

38 See the quotation of Angerer in Maehle, 2000, p. 207.
39 *Der Chirurg,* founded and edited by Martin Kirschner, 1929ff.
40 Guleke, 1937, pp. 359–81.
41 Ibid., pp. 268f.
42 Quotation from Hausberg, 1906, p. 85.

References

Ackerknecht, Erwin H. (1953), *Rudolf Virchow. Doctor, Statesman and Anthropologist*, University of Wisconsin Press, Madison/W.

Aschoff, L. (1925), 'Über die Bedeutung der Leichenöffnungen und des Tierexperiments für die Volksgesundheit und die soziale Wohlfahrtspflege', *Wissenschaft und Werktätiges Volk*, pp. 151–85.

Aschoff, L. (1966), *Ein Gelehrtenleben in Briefen an die Familie*, Schulz, Freiburg i. Br.

Bächtold, H. (1917), *Deutscher Soldatenbrauch und Soldatenglaube*, Trübner, Strasbourg.

Bauer, H.G. (1929), *Die Behandlung des menschlichen Leichnams im geltenden deutschen Recht*, Große, Dresden.

Benaroyo, L. (1998), 'Pathology and the Crisis of German Medicine (1920–1930): A Study of Ludwig Aschoff's Case', in Prüll, C.-R. (ed.), *Pathology in the 19th and 20th Centuries. The Relationship between Theory and Practice* (EAHMH, Network Series, vol. 2), EAHMH Publications, Sheffield, pp. 101–13.

Böckle, F. (1983), 'Pietät oder Nächstenliebe? Zur sittlichen Bewertung der medizinischen Obduktion', *Der Pathologe*, vol. 4, pp. 1–2.

Bohne, G. (1932), 'Das Recht zur klinischen Leichensektion', in *Festgabe für Richard Schmidt. Zu seinem siebzigsten Geburtstag am 19. Januar 1932 überreicht von Verehrern und Schülern*, Hirschfeld, Leipzig, pp. 105–76.

Brugger, C.M. and Kühn, H. (1979), *Sektion der menschlichen Leiche. Zur Entwicklung des Obduktionswesens aus medizinischer und rechtlicher Sicht*, Enke, Stuttgart.

Büchner, F. (1957), 'Ludwig Aschoff', in Vincke, J. (ed.), *Freiburger Professoren des 19. und 20. Jahrhunderts*, Albert, Freiburg i.Br., pp. 11–20.

Büchner, F. (1943), 'Gedenkrede auf Ludwig Aschoff, gehalten bei der Gedenkfeier der Universität Freiburg am 5. Dezember 1943', *Feldpostbrief der Medizinischen Fakultät der Universität Freiburg/Brsg.*, no. 4, pp. 1–24.

Buscher, D. (1980), *Die wissenschaftstheoretischen, medizinhistorischen und zeitkritischen Arbeiten von Ludwig Aschoff*, MD thesis, Freiburg i. Br.

Doerr, W. (1979), 'Geleitwort', in Brugger, C.M. and Kühn, H., *Sektion der menschlichen Leiche. Zur Entwicklung des Obduktionswesens aus medizinischer und rechtlicher Sicht*, Enke, Stuttgart, pp. vii–viii.

Elkeles, B. (1996), *Der moralische Diskurs über das medizinische Menschenexperiment im 19. Jahrhundert*, Fischer, Stuttgart.

Eser, A. (1990), 'Beobachtungen zum "Weg der Forschung" im Recht der Medizin', in idem (ed.), *Recht und Medizin*, Wiss. Buchgesellschaft, Darmstadt, pp. 1–42.

Eulner, H.-H. (1970), *Die Entwicklung der medizinischen Spezialfächer an den Universitäten des deutschen Sprachgebietes* (Studien zur Medizingeschichte des 19. Jahrhunderts, vol. 4), Enke, Stuttgart.

Fischer, W. (1964), 'Ludwig Aschoff. 1866–1942', in Freund, H. and Berg, A. (eds), *Geschichte der Mikroskopie. Leben und Werk grosser Forscher*, vol. 2, Umschau, Frankfurt/M., pp. 13–21.

Fuchs, F. and Panek, O. (1928), 'Statistische und klinische Beobachtungen über die an der Klinik Hochenegg 1904–1926 behandelten inoperablen Krebsfälle', *Archiv für klinische Chirurgie*, vol. 151, p. 263.

Geiger, P. (1932–33), 'Leiche', in Bächthold-Stäubli, H. et al., *Handwörterbuch zur deutschen Volkskunde*, Abt.1: Aberglaube, vols 1–10, Berlin (1927–42), vol. 5, de Gruyter & Co., Berlin and Leipzig, col. 1024–60.

Geiger, P. (1932–33), 'Leichenschändung', in Bächthold-Stäubli, H. et al., *Handwörterbuch zur deutschen Volkskunde*, Abt.1: Aberglaube, vols 1–10, Berlin (1927–42), vol. 5, de Gruyter & Co., Berlin and Leipzig, col. 1093f.

Geiger, P. (1936–37), 'Tote', in Bächthold-Stäubli, H. et al., *Handwörterbuch zur deutschen Volkskunde*, Abt.1: Aberglaube, vols 1–10, Berlin (1927–42), vol. 8, de Gruyter & Co., Berlin and Leipzig, col. 1019–34.

Göckenjan, G. (1989), 'Wandlungen im Selbstbild des Arztes seit dem 19. Jahrhundert', in Labisch, A. and Spree, R. (eds), *Medizinische Deutungsmacht im sozialen Wandel*, Psychiatrie-Verlag, Bonn, pp. 89–102.

Grynaeus, T. (1993), 'Weiterleben der heilenden Volksbräuche und -glauben in einem neugesiedelten Dorf in Ungarn', *Curare*, vols 3–4, pp. 191f.

Guleke, N. (1937), 'Die Grenzen der Verantwortlichkeit des Chirurgen', *Verhandlungen der Deutschen Gesellschaft für Chirurgie*, pp. 359–81.

Häßner, H. (1916), 'Pathologische Anatomie im Felde', *Virchows Archiv für pathologische Anatomie und Physiologie und für klinische Medizin*, vol. 221, pp. 309–10.

Hausberg (1906), 'Abänderungen des deutschen Strafgesetzbuches in Bezug auf die Heilkunde', *Ärztliches Vereinsblatt*, vol. 33, pp. 78–93.

Heym, G. (1984), 'Die Sektion', in Rölleke, H. (ed.), *Georg Heym Lesebuch: Gedichte, Prosa, Träume, Tagebücher*, Beck, Munich, pp. 181–3.

Klasen, E.-M. (1984), *Die Diskussion über eine 'Krise' in der Medizin in Deutschland zwischen 1925 und 1935*, MD thesis, Mainz.

Kocher, T. (1907), *Chirurgische Operationslehre*, Fischer, Jena.

König, F. (1931), 'Etwas von Leib und Seele des Chirurgen; zu gleich ein Stück chirurgischer Ethik', *Deutsche Zeitschrift für Chirurgie*, vol. 234, p. 25.

Krecke, A. (1913), *Beiträge zur praktischen Chirurgie 1910–12*, Lehmann, Munich, pp. 713–18.

Krecke, A. (1926), *Beiträge zur praktischen Chirurgie 1923–26*, Lehmann, Munich, pp. 1–22.

Kronfeld, E.M. (1915), *Der Krieg im Aberglauben und Volksglauben. Kulturhistorische Beiträge*, Schmidt, Munich.

Laufs, A. (1986), 'Arzt und Recht im Wandel der Zeit', *Medizinrecht*, vol. 4, pp. 163–70.

Liek, E. (1926), *Der Arzt und seine Sendung. Gedanken eines Ketzers*, Lehmanns, Danzig.

Löffler, P. (1975), *Studien zum Totenbrauchtum in den Gilden, Bruderschaften und Nachbarschaften Westfalens vom Ende des 15. bis zum Ende des 19. Jahrhunderts* (Forschungen zur Volkskunde, vol. 47), Regensberg, Münster.

Lubarsch, O. (1931), *Ein bewegtes Gelehrtenleben. Erinnerungen und Erlebnisse. Kämpfe und Gedanken*, Springer, Berlin.

Maehle, A.-H. (1999), 'Professional Ethics and Discipline: The Prussian Medical Courts of Honour, 1899–1920', *Medizinhistorisches Journal*, vol. 34, pp. 309–38.

Maehle, A.-H. (2000), 'Assault and Battery, or Legitimate Treatment? German Legal Debates on the Status of Medical Interventions without Consent, c. 1890–1914', *Gesnerus*, vol. 57, pp. 206–21.

Maio, G. (1996), 'Das Humanexperiment vor und nach Nürnberg: Überlegungen zum Menschenversuch und zum Einwilligungsbegriff in der französischen Diskussion des 19. und 20. Jahrhunderts', in Wiesemann, C. and Frewer, A. (eds), *Medizin und Ethik im Zeichen von Auschwitz. 50 Jahre Nürnberger Ärzteprozeß*, Palm and Enke, Erlangen and Jena, pp. 45–78.

Maulitz, R.C. (1993), 'The Pathological Tradition', in Bynum, W.F. and Porter, R. (eds), *Companion Encyclopedia of the History of Medicine*, vol. 1, Routledge, London and New York, pp. 169–91.

Meeßen, R. (1975), *Die Freiburger Pathologie, ihre Entstehung und Fortentwicklung*, MD thesis, Freiburg i. Br.

Oberhoff, G. (1935), *Über die Rechtswidrigkeit und Strafbarkeit klinischer Leichensektionen*, Law thesis, Lechte, Erlangen.

Orth, J. (1906), 'Das Pathologische Institut zu Berlin', *Berliner Klinische Wochenschrift*, vol. 43, pp. 817–26.

Pantel, J. and Bauer, A. (1990), 'Die Institutionalisierung der pathologischen Anatomie im 19. Jahrhundert an den Universitäten Deutschlands, der deutschen Schweiz und Österreichs', *Gesnerus*, vol. 47, pp. 303–28.

Probst, C. (1994), 'Die Religiosität des Landvolks im Urteil der Ärzte. Aus den Landes- und Volksbeschreibungen der bayerischen Amtsärzte um 1860', *Die Medizinische Welt*, vol. 45, pp. 152–6.

Prüll, C.-R. (1995), 'Aschoff, Ludwig', in Eckart, W.U. and Gradmann, C. (eds), *Ärzte-Lexikon. Von der Antike bis zum 20. Jahrhundert*, Beck, Munich, pp. 24f.

Prüll, C.-R. (1996), 'Die Sektion als letzter Dienst am Vaterland. Die deutsche "Kriegspathologie" im Ersten Weltkrieg', in Eckart, W.U. and Gradmann, C. (eds), *Die Medizin und der Erste Weltkrieg* (Neuere Medizin- und Wissenschaftsgeschichte. Quellen und Studien, vol. 3), Centaurus, Pfaffenweiler, pp. 155–82.

Prüll, C.-R. (2002), *Medizin am Toten oder am Lebenden? – Pathologie in Berlin und in London 1900–1945*, Schwabe, Basle, forthcoming.

Richardson, R. (1988), *Death, Dissection and the Destitute*, Penguin, London.

Ricker, G. (1921), 'Die pathologische Anatomie der frischen mechanischen Kriegsschädigungen des Hirnes und seiner Hüllen', in Aschoff, L. (ed.), *Pathologische Anatomie* (Handbuch der Ärztlichen Erfahrungen im Weltkriege 1914–18, ed. by O. von Schjerning, vol. 8), Barth, Leipzig, pp. 334–83.

Rössle, R. (1930), 'Allgemeine Pathologie und pathologische Anatomie in ihren gegenseitigen Beziehungen', *Münchener Medizinische Wochenschrift*, vol. 77, pp. 3–15.

Roßner, H.-J. (1998), *Begrenzung der Aufklärungspflicht des Arztes bei Kollision mit anderen ärztlichen Pflichten. Eine medizinrechtliche Studie mit vergleichenden Betrachtungen des nordamerikanischen Rechts* (Medizin und Recht, vol. 40), Lang, Frankfurt/M., pp. 46–53.

Sauerteig, L. (2000), 'Ethische Richtlinien, Patientenrechte und ärztliches Verhalten bei der Arzneimittelerprobung (1892–1931)', *Medizinhistorisches Journal*, vol. 35, pp. 303–34.

Schepper-Lambers, F. (1992), *Beerdigungen und Friedhöfe im 19. Jahrhundert in Münster* (Beiträge zur Volkskultur in Nordwestdeutschland, vol. 73), Coppenrath, Münster.

Schmiedebach, H.-P. (1989), 'Der wahre Arzt und das Wunder der Heilkunde. Erwin Lieks ärztlich-heilkundliche Ganzheitsideen', in *Der ganze Mensch und die Medizin* (Argument Sonderband 162), Argument, Hamburg, pp. 33–53.

Seesemann, H. (1928), 'Ist eine heimliche Leichensektion strafbar?', *Ärztliches Vereinsblatt*, vol. 55, p. 721.

Seesemann, H. (1931), 'Ist eine heimliche Leichensektion strafbar?', *Ärztliches Vereinsblatt*, vol. 60, pp. 45f.

Seidler, E. (1976), 'Pathologie in Freiburg', *Beiträge zur Allgemeinen Pathologie und pathologischen Anatomie*, vol. 158, pp. 9–22.

Seidler, E. (1991), *Die Medizinische Fakultät der Albert-Ludwigs-Universität Freiburg im Breisgau. Grundlagen und Entwicklungen*, Springer, Berlin, Heidelberg etc.

Tröhler, U. (1993), 'Surgery (modern)', in Bynum, W.F. and Porter, R. (eds), *Companion Encyclopedia of the History of Medicine*, vol. 2, Routledge, London and New York, pp. 984–1028.

Übelhack, B. (2002), *Ärztliche Ethik – Eine Frage der Ehre ? Die Urteile der ärztlichen Ehrengerichtshöfe in Preußen und Sachsen 1918-1933*, Lang, Frankfurt/M., in press.

Virchow, R. (1886), 'Das Pathologische Institut', in Guttstadt, A. (ed.), *Die naturwissenschaftlichen und medicinischen Staatsanstalten Berlins. Festschrift für die 59. Versammlung deutscher Naturforscher und Aerzte*, Hirschwald, Berlin, pp. 288–300.

Weindling, P. (2001), 'The Origins of Informed Consent: The International Scientific Commission on Medical War Crimes, and the Nuremberg Code', *Bulletin of the History of Medicine*, vol. 75, pp. 37–71.

Winau, R. (1996), 'Medizin und Menschenversuch. Zur Geschichte des "informed consent"', in Wiesemann, C. and Frewer, A. (eds), *Medizin und Ethik im Zeichen von Auschwitz. 50 Jahre Nürnberger Ärzteprozeß*, Palm & Enke, Erlangen and Jena, pp. 13–29.

Chapter 5

Human Research: From Ethos to Law, from National to International Regulations

Ulrich Tröhler

Introduction

Familiar to antiquity, human experimentation emerged again in the seventeenth century and became more prevalent after the middle of the eighteenth century (Howard-Jones, 1982; Bynum, 1988). The reasons for this were manifold, rooted in medicine, culture and science. Within medicine the discovery of the circulation of the blood in the first half of the seventeenth century led to further physiological investigations and attempts at intravenously applying traditional drugs. The (side) effects of new drugs were also tested in various ways in the eighteenth century (Maehle, 1999a). New surgical procedures, too, were evaluated comparatively (Tröhler, 2000a). Finally there was the inoculation of smallpox – constituting, in fact, a series of experiments in preventive medicine carried out throughout Europe. However, in this period, testing (new) medical interventions such as drugs, using new surgical procedures or preventive measures and evaluating non-therapeutic studies, i.e. (patho-)physiological experiments that aimed at understanding bodily functions in health and disease, were not differentiated categorically.

This chapter examines ethical arguments related to human research, looking at why, when and where they were formulated and where regulating practices emerged. It also briefly considers the question of researchers' compliance with such regulations. My emphasis will be on the emergence of international codes of ethics.

'For the Good of Mankind': Little Need for Formal Regulations Prior to World War II

Under the *ancien régime* doctors had little difficulty finding participants for various kinds of human experimentation, provided they were in a position simply to order patients to comply or, alternatively, to pay them to undergo such procedures, even if they did not understand them. Thus, in the 1660s, the effect of a transfusion of the 'good' blood of a lamb (whose 'bad blood' was evacuated by copious venesection) was tested on an insane patient in Paris who was paid for volunteering. This test caused moral concerns to be raised, due partly to its unforeseen effects, which proved to be life-threatening. Such experimentation was subsequently forbidden by the Paris parliament (Starr, 1998). In the second half of the eighteenth century new ways of amputating limbs, of treating wounds, and of operating on cataracts were tried on military pensioners and on the battlefield. The effectiveness of various antiscorbutics and drugs against different types of fevers were tested on sailors by comparing them with the traditional treatments.

These eighteenth century clinical trials raised ethical problems subsequently associated with well-controlled, published human experimentation. Often enough this did not apply to haphazard trial and error actions, frequent enough in daily practice. Some army surgeons saw no problem in allocating soldiers to groups during a battle, for instance, to find out whether the mortality of amputation was lower immediately after the injury or if the intervention was delayed for a few days. Rational arguments existed for both methods. Others hesitated.

When it came to carrying out a study of the benefits of delayed versus immediate amputation during the Napoleonic Wars, the British Army Surgeon George James Guthrie had scruples about surgery after the 'success' he deemed to have seen with immediate intervention. He felt himself not 'authorised to commit murder for the sake of experiment' (Guthrie, 1815: 39). He preferred to rely on a retrospective analysis of his casebooks instead, despite a theoretical insight into the necessity of conducting prospective comparative trials and the unique opportunities a commanding military surgeon had to enforce them. Another author, Charles McLean, however, realised in 1818 the ethical double-standard involved in this pretended 'reluctance to try experiments with the lives of men … as if the practice of medicine, in its conjectural state, were anything else, than a continued series of experiments, upon the lives of our fellow creatures' (McLean, 1817–18, vol. 2: 500–4). This reflection on the morality of acting in the light of evidence gathered in traditional ways remained an isolated one, however, while Guthrie's argument represented the widespread opinion that those participating in research themselves 'should benefit from the trials to which they were subjected and that they must not be put in danger for the sake of scientific curiosity' (Maehle, 1999a: 268–9).

Indeed, 50 years earlier, James Lind had reacted in exactly that way, but the case ended quite differently. When starting a trial of the 'malt-wort' ordered by the Admiralty, Lind acknowledged the 'murmur and disgust' after withholding vegetables from scurvy patients at Haslar (because of the belief that this would improve their condition) and stopped it. But the Admiralty ordered it to be taken up at sea 'where it was expected that patients would cheerfully submit' (Macbride, 1764: 174–5). Although there was a climate of 'ethical awareness', in that some research minded doctors felt the dilemma between patient care and the advancement of knowledge, giving relevant information and obtaining consent were obviously not a major issue. Hierarchical power was central in this and other trials, particularly in the Navy, but also in the Army and probably in civilian institutions.

There was another way to circumvent the harm potentially present in the use of any new method. Because he felt 'it would be unjustifiable to neglect for the sake of experiment any means of safety', James Currie in 1804 'superadded' his cold water bathing to the traditional fever treatment – in other words, he used today's 'add-on design' (Currie, 1804: 408). This implied, of course, that he considered traditional bleeding safe (on what evidence, we do not know). These discussions about risk, even sacrifice 'for the sake of experiment', and of safety, illustrate the ambiguity of the notion of experiment. Many eighteenth century doctors understood its everyday meaning, that is, a straightforward test with unknown (yet sometimes hoped for) beneficial results. For some this meant a planned intervention under well controlled conditions and circumstances with respect to the selection of patients, the treatment(s) given *and* the particular care with which the patients were attended. Finally, Charles McLean held routine clinical practice based on 'conjectural', in other words, inferior, evidence as nothing other than an uncontrolled experiment, an important statement which has often enough not been understood by either doctors or the public (Tröhler, 2000).

These issues, arguments and possible solutions continued to be advanced throughout the nineteenth century when the ethos of increasing scientific knowledge as a basis for progress also prevailed in medicine. But there were still others. Further safeguards against inflicting damage to humans 'for the sake of experiment' were formulated, such as the requirement of previous animal studies and/or self-experimentation by doctors, and there is evidence that these principles were actually followed (Tröhler and Maehle, 1987). Animal experiments had been used since the seventeenth century, although the transferability of results to humans was sometimes questioned (Maehle and Tröhler, 1987). Nineteenth century surgery again offered many examples of this strategy, such as trying new techniques of ovarectomy and nephrectomy as well as thyroidectomy on animals and observing their consequences before operating on patients (Tröhler, 1993). Auto-experimentation played a role in the history of inhalation anaesthesia. Because of the prevailing ethos and/or doctor and patient insistence on surgery as a 'last hope', however, other 'first operations' were performed directly on patients. A case in point was the first successful heart suture undertaken in 1896 by Ludwig Rehn of Frankfurt in an emergency situation (Tröhler, 1998). After the success of vaccination against smallpox and with the rise of microbiology, preventive and therapeutic measures against other infectious diseases, too, were tried throughout the nineteenth century, some directly on patients because progress seemed more tangible. This was the case for both Louis Pasteur's anti-rabies serum and his German competitor Robert Koch's tuberculosis treatment with *Tuberkulin* (Geison, 1995; Gradmann, 2000). Pasteur's treatment ultimately proved a success; Koch's rapidly became a failure.

Clearly, there were no acknowledged rules for the type of evidence seen as sufficient for an innovation to be considered safe enough for general practice. This held true for modern surgery as well as for new drugs, chemical and others, produced by an expanding pharmaceutical industry. No licensing body existed. Neither was there a set of rules on the ethics of generating scientific evidence (Sauerteig, 2000). While the need for planned, well controlled intervention was stressed as indispensable for the advance of medicine in the many editions and translations of Claude Bernard's *Introduction à la médecine expérimentale* (1865), this and texts of the same kind had no explicit headings on the ethics of (human) research. Instead, Bernard reaffirmed the traditional caution of 'never performing on man an experiment which might be harmful to him to any extent, even though the result might be highly advantageous to science, i.e. to the health of others' (Bernard, 1949: 101). He held that 'many physicians attack experimentation believing that medicine should be a science of observation'. But, he pointed out, as McLean had done 50 years before him, that 'physicians make therapeutic experiments daily on their patients, so this inconsistency cannot stand careful thought. Medicine by nature is an experimental science, but it must apply the experimental method systematically' (Bernard, 1949: 18). On the other hand, the eighteenth century British advocates of professional medical ethics, John Gregory and Thomas Percival, had in fact dealt with ethical issues inherent in therapeutic tests as a form of human experimentation, the former quite extensively (McCullough, 1998), the latter simply insisting on their usefulness 'for the public good', provided they were scrupulously and conscientiously carried out and passed a process of prior peer review (Tröhler, 2000: 130; Baker, 1993). It was left to the responsibility and the individual doctor's conscience to judge what that meant. The idea and meaning of giving information and obtaining consent were not heeded (Rothman, 1998) and the whole subject became conspicuous by its absence from the subsequent British and German deontological literature up to 1930 – with one exception. Nor was hardly any misdemeanour in this area dealt with by the professional courts of honour of these two countries, although cases of what we would today call 'severe abuse' had occurred (Maehle, 1999b; Übelhack, 2002; Smith, 1994). The exception was Albert Moll of Berlin. As a practising neurologist he was interested in medicine related to sexual diseases and sexual reform. He was certainly not representative of mainstream medicine. In his 650-page *Ärztliche Ethik* (1902), he listed about 600 cases of non-therapeutic research, published in the medical press without further comment, that he considered unethical because of the damage inflicted, unclear or evidently useless application, bad design and – most notably – absence of any form of information and/or consent by the patient-subjects (Moll, 1902).

The notion of informing a patient and obtaining consent did indeed exist but in the legal rather than the medical arena (Maehle, 2000). This holds particularly true for English law, which as early as 1830 was interpreted as obliging the physician to obtain the informed consent of a potential participant in experiments. 'Otherwise, [the doctor] would be obliged to provide compensation for any injury that might arise from adopting a new method of treatment' (Perley et al., 1992: 150). In the middle of the nineteenth century there were isolated criminal trials in France and Germany, and in the 1880s and 1890s in Norway and Austria, respectively. Doctors were condemned – mildly – for omitting to inform and not seeking consent. In the view of judges they had therefore inflicted physical injury on patients undergoing experimentation. Formal links between the doctors' ethics and administrative and legal practice were only established afterwards, in 1899. Albert Neisser, professor of dermatology and venereology at the University of Breslau (and discoverer of the gonorrhoea bacillus), was condemned by the Royal Prussian Disciplinary Court for his actions. This was a tribunal concerned exclusively with the civil service, i.e. neither an ordinary criminal court nor a professional court of honour.

Neisser had injected cell-free serum from syphilitic patients into eight patients, some being minors and others prostitutes, without informing them nor obtaining their consent. He wanted to test whether this might provide immunity against contracting syphilis. This and similar cases were exposed by the liberal press, which saw abuse inflicted on the poor in favour of finding a cure to mask the double standards of the wealthy. This raised public and even heated parliamentary debates, causing widespread consultations with medical and legal authorities by the Minister in charge. After the Neisser trial, official administrative regulations were introduced.

In December 1900 the Prussian Minister of Religious, Educational and Medical Affairs issued specific *Directives* (Minister, 1901). They were addressed to all heads of state clinics, polyclinics and other hospitals in the country. The physicians and surgeons-in-chief were advised that intervention with other than diagnostic and therapeutic aims or for the purpose of immunisation was excluded under all circumstances in minors and other legally not fully competent persons, and unless the potential participant had consented unambiguously after explanation of possible adverse consequences. These conditions and the precise circumstances of the study had to be documented in the case notes. This administrative directive was the first explicit, albeit legally weak, regulation of human experimentation. But, in actual fact, these requirements concerned only non-therapeutic research, designed to advance (patho-)physiological knowledge. Even the frequent immunisations then prevalent were not considered: it was taken for granted that 'patients in a public hospital submitted regularly ... to new methods of treatment and diagnostic experiments' (quoted by Elkeles, 1996: 209). In 1906 the Austrian Ministry of Education issued a nearly identical directive (Elkeles, 1996). Paul Ehrlich, when introducing his Salvarsan, the world's first chemotherapeutic agent, was certainly aware of the 1900 *Directives* and acted cautiously to avoid public scandal (Sauerteig, 2000).

Concerns and debates continued in interwar Germany (Sauerteig, 2000). A generation later, in 1931, the 1900 *Directives* were further elaborated as *Guidelines for Novel Therapeutic Trials and for Performing Scientific Experiments in Humans* (Winau and Vollmann, 1996). Issued by the Reich's Minister of the Interior, the *Guidelines* included, as indicated by the title, both therapeutic and non-therapeutic research. As the *Directives* had done, the *Guidelines* required giving information and obtaining consent as well as documenting these procedures. As administrative measures, they had no standing in criminal law. The question of the wider practical application of these two regulatory measures remains unresolved (Elkeles, 1996). Although never revoked, they were certainly ignored during National Socialism, particularly during the Second World War.

In the USA as well, before World War II, some doctors were aware of the moral dimensions of human research, the possible conflict between scientific advance and potential harm to patients, and the responsibility of physicians for their patients (Lederer, 1995).

'Never Again': Regulations in the Post-World War II Phase

As a reaction to the international public outcry over the atrocities of German researchers, which were made public at the end of the Second World War, the *Nuremberg Code* was drawn up during the trial against Nazi medical war criminals in 1946–1947 by an American jury. In ten principles the *Code* specified the prerequisites that must be met for human experiments to be morally acceptable (Annas and Grodin, 1992). The code could be interpreted as emphasising non-therapeutic studies only, as it did not mention specific types of human research, but just used the term 'experiment' throughout (Deutsch, 1997). In terms of contents, 12 'markers' can be identified (Herranz, 1998). They are outlined in Table 5.1. Since the Nuremberg Trial was based on several official declarations by the Allies regarding the post-war prosecution of war crimes (e.g., London 1942, Moscow 1943), this *Code*, in fact, constituted a document of international public law (Arnold and Sprumont, 1998). There was, however, no corresponding international institution to control its application or to prosecute violations. Apart from attempts by the US

Table 5.1 Markers of the Nuremberg Code

MARKERS ADOPTED: Criteria from the Nuremberg Code that were incorporated into the Declaration of Helsinki.

1 Voluntary consent of the subject.
2 An expected beneficial outcome of the experiment.
3 Prior experimentation on animals.
4 Avoidance of unnecessary pain and harm.
5 Avoidance of any risks of death or disablement.
6 The risks taken must not exceed the expected advantages.
7 Protecting against the possibility, however slight, of injury, disablement, or death.
8 Scientifically and technically qualified experimenters.
9 The subject's freedom to retract consent.
10 The experimenter's obligation to stop the experiment if it turns out that it is dangerous.

MARKERS ABANDONED: Criteria from the Nuremberg Code that were not incorporated into the Declaration of Helsinki.

11 The personal, non-transferable responsibility of everyone involved in initiating and carrying out the experiment to assure the ethical quality of the subject's consent.
12 The right of the subject to quit the experiment if his condition seems to rule out the possibility of continuing.

government and the World Medical Association (WMA, see below) to implement it in military and clinical contexts (Baker, 1998), it probably had no great practical significance for some two decades (Herranz, 1998). Yet with hindsight we can consider the code of Nuremberg in view of its contents – particularly its insistence on informed consent, on balancing potential harm and benefit, and on the right of the participant to quit the experiment – and because of its legal status, the first blueprint designed in international public law with reference to medicine and human rights, on which others expanded significantly in the decades to come (Schuster, 1997; Drinan, 1992). At the same time, however, the *Code*'s professional origin outwith medicine was a hindrance for its acceptance by the *medical* research community.

An international document clearly showing the influence of the *Nuremberg Code* was the four part *Conventions of Geneva* of August 1949, put forward under the auspices of the Red Cross. They aimed at establishing 'a humanitarian law of armed conflict which aims at protection of non-combatant military personnel and civilians not involved in the hostilities' (Bassiouni et al., 1981: 1659). Each of the four *Conventions* had an article prohibiting criminal biological tests and medical experimentation on civilian and captive wounded or sick persons. The conventions were subsequently updated and provided with 'Additional Protocols' in 1977 (Perley et al., 1997).

The origins of another international document, the *International Covenant on Civil and Political Rights*, were also significantly influenced by the *Nuremberg Code*. Although it was adopted by the United Nations General Assembly in 1966 and took legal effect only in 1976, its elaboration began in 1947 in the earliest sessions of the UN Commission of Human Rights. The first draft already contained an article concerning human research which, in the course of time, followed ever more closely the wording of the *Code* (Perley et al., 1997).

These two documents outline the complementary development of ethics codes with their traditional insistence on the duties of scientists on the one hand and modern statements on the rights of participants on the other.

The Nuremberg trial had publicly exposed abuses of modern medicine. It had shown that the medical profession lacked a written *charta* professing its traditional ethos and fostering a vision of the ideal doctor, who aimed at doing good, preventing harm and who respected confidentiality. This was consistently taken for granted and deemed to be sufficient. It was in this state of affairs that the WMA was founded in Geneva in 1947. It immediately became very active with respect to morality in medicine. As early as 1948, it issued the *Declaration of Geneva*, a modern version of the Hippocratic Oath, and a year later an *International Code of Medical Ethics*. Both contained passages with a bearing on human research. The *Declaration* stipulated that 'even under threat ... I will not use my medical knowledge contrary to the laws of humanity' – on which the *Nuremberg Code* had been based. The *International Code* warned against damaging patients' interests when 'providing medical care which might have the effect of weakening the physical and mental condition of the patient', for instance when doing so without therapeutic necessity in physiological studies (quoted from Reich, 1995, vol. 5, p. 2647).

A need for regulation was also recognised in Britain, where the Medical Research Council issued a memorandum on the ethics of clinical trials in 1953. This memorandum was widely circulated among 'medical members of the [MRC] staff and research workers in receipt of grants ..., the Deans of the Medical Faculties and Medical Schools, the Secretaries of the Medical Scientific Societies and the Editors of the relevant scientific journals' (Medical Research Council, 1953).

In 1954 the WMA issued specific *Principles for Those in Research and Experimentation*, designed to implement the code of Nuremberg and to allow research under certain conditions (Reich, 1995, vol. 5, p. 2764). Implicitly forbidden by the *Nuremberg Code* was research on unconscious and psychiatric patients and on children. But this could be circumvented, for instance, by introducing the notion of surrogate consent (Baker, 1998).

While the Public Health Council of the Netherlands tentatively adopted them, massive resistance in the research community prevented the implementation of these *Principles*, even in the United States and in Germany. Elsewhere they were seen as an unnecessary hindrance to research and progress, because abuses were attributed to (Nazi) criminals, not the medical profession in general (Baker, 1998; Winslade and Krause, 1998). The *Principles* are practically forgotten today: S. Fluss does not mention them in his comprehensive listing of international guidelines on bioethics (Fluss, 1999). However, these documents of the 1950s initiated a movement towards the codification of ethical standards in various health care fields.

The Wave of Codification of Ethics: A New Phenomenon

Around 1960, due to public scandals, human research again came under scrutiny. The recognition that new drugs might have rare but devastating effects, such as the limb abnormalities caused by thalidomide, made regulation more urgent. Thalidomide made international headlines. Local and national bodies such as the Harvard Medical School (1961), the British Medical Association and the British Medical Research Council (1963) reacted by issuing appropriate ethical guidelines. Academic and professional organisations of other European countries, such as Germany and Switzerland, followed suit, although with some delay. Indeed, in the 1960s many national governments passed formal acts which tightened up or introduced the licensing of new pharmaceutical products. Such was the case in the United States (Glantz, 1992; Baker, 1997), while the issue was avoided in France (Maio, 2001). On an international level, in 1964, the WMA *Principles* were reformulated under the title of *Recommendations Guiding Physicians in Biomedical Research involving Subjects*. This document, now known as the WMA *Declaration of Helsinki* (I), was an example of internationalising the already ongoing national regulation of moral issues arising in health care fields. Both inter-governmental and non-governmental organisations (IGOs and NGOs, respectively) authored such documents. The latter did so partly in order to prevent 'the enactment of criminal or legislative measures' in a field which doctors clearly considered their own province (Arnold and Sprumont, 1998).

This phenomenon is illustrated in Figure 5.1. It shows the international guidelines, declarations, recommendations, etc., here summarised by the term 'Ethics Codes', issued annually between 1947 and 2000. So far the total number, as listed by S. Fluss, amounts to 326 variants. Eleven IGOs, ranging from the United Nations (1966) and its suborganisations (UNESCO, WHO, UNAIDS), via, for example, the Council of Europe, the European Commission, the Organisation of African Unity, to the World Labour Organisation (1999), have issued 107 such codes.

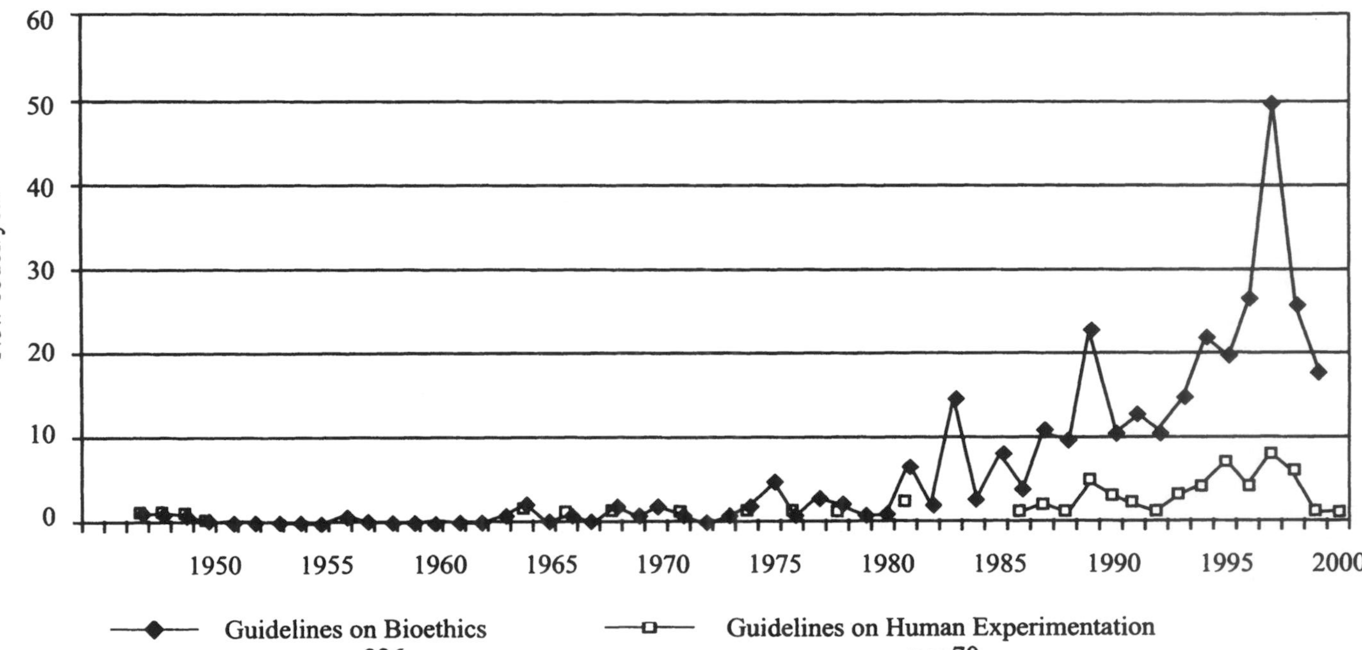

Note: The number of international ethics codes (i.e., guidelines, recommendations, resolutions, conventions) issued by Inter-Governmental (IGO) and Non-Governmental Organisations (NGO) as well as miscellaneous international texts published each year (from 1947 to 2000) is represented. Each amended version of a given code was counted individually. Drawn from data collected by S. Fluss, 1999, and my own research.

Figure 5.1 International ethics codes 1947–2000

Only 36 NGOs, ranging from the WMA (1948), via – to name just a few – Amnesty International, the Council for International Organisations of Medical Sciences (CIOMS), the European Forum of Good Clinical Practice (EFGCP), the International Council of Nurses, the Human Genome Organization, to the International Bar Association (2000), have produced 186 documents; the governmental entity of the Vatican, six, and miscellaneous bodies, more than 20. The number of national codes can only be guessed at. The Swiss Academy of Medical Sciences alone, between 1969 and Spring 2001, published 27 versions in 13 fields, including five versions of two codes regarding research on human beings (Tröhler, 1999; SAMW, 2001).

This chapter cannot analyse the contents of such an enormous amount of source material. (For an analysis respecting ethical issues, see e.g. Sass, 1988 and Veatch, 1995.) Rather, I will suggest some of the reasons for this new phenomenon. Internal, scientific, as well as external, socio-cultural reasons, play a part. Taboos were being broken in specific fields. In reproductive medicine, for instance, Louise Brown, the first apparently healthy baby stemming from in vitro fertilisation, was born in the UK in 1978. In 1985, for the first time in history, a pregnancy was brought to full term by a surrogate mother; in 1990 the ova of aborted foetuses were marketed; 1998 featured, still in the UK, the successful cloning of the sheep Dolly, and in 2000 Parliament in Westminster discussed human 'therapeutic' cloning (and approved it). In the early 1960s, euthanasia became an issue in many countries because of the availability of life-sustaining technologies. These were also a prerequisite for using the criterion of brain death, which was sometimes discussed together with organ transplantation routines (Schöne-Seifert, 1999). Most recently xenotransplantation and tissue engineering raised moral issues. Modern genetics, too, question fundamental concepts of society as well as of health care policies.

Since no progress in these areas is possible without human experimentation, one-fifth of all international codes have dealt with issues related to this major single field of concern in the aftermath of Nuremberg. More and more NGOs and IGOs have become active in related moral concerns in the 1990s (Figure 5.2). Further scandals in research showed the inadequacy of the principle of subsidiarity, particularly in Germany and the United States (Baker, 1998). National governments have intervened increasingly and in various ways in a field which had previously been regulated within the medical community (Winslade and Krause, 1998). France was the first country, in 1983, to establish, by presidential decree, a national bioethics committee as a consultation body (*Comité Consultatif National d'Ethique de la Médecine et des Sciences de la Vie*). It proposed moral norms and practical recommendations which 'contaminated' subsequent laws, e.g. the Laws of Bioethics (*lois de bioéthique*) of 1994, including a regulation of human experimentation (Mathieu, 1998). This political process did not fail to become internationalised, and the IGOs'

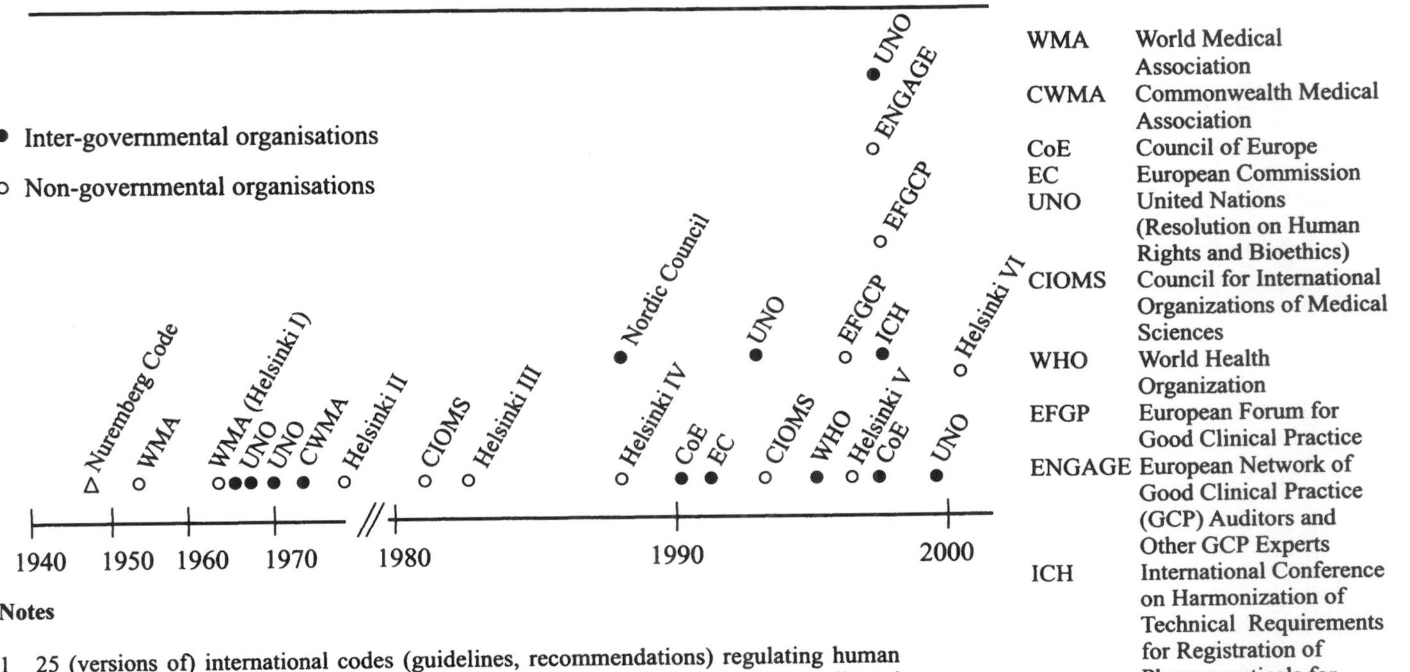

Abbreviations

WMA	World Medical Association
CWMA	Commonwealth Medical Association
CoE	Council of Europe
EC	European Commission
UNO	United Nations (Resolution on Human Rights and Bioethics)
CIOMS	Council for International Organizations of Medical Sciences
WHO	World Health Organization
EFGP	European Forum for Good Clinical Practice
ENGAGE	European Network of Good Clinical Practice (GCP) Auditors and Other GCP Experts
ICH	International Conference on Harmonization of Technical Requirements for Registration of Pharmaceuticals for Human Use

Notes

1 25 (versions of) international codes (guidelines, recommendations) regulating human experimentation are listed in order of the dates of publication. Drawn from data collected by Fluss, 1999, and my own research.

2 The UN codes (Resolutions) deal only in part with human experimentation. Further explanations are contained in the text.

Figure 5.2 International ethics codes on human experimentation

involvement reflects the growing attention of international legislative bodies. But had this not been the intention of the *Nuremberg Code* 40 years earlier?

For example, the Council of Europe, and later the European Commission (EC) have developed activities with respect to research on human beings (Figure 5.2). It is a little known fact that there is a European Commissioner of Health and Consumer Protection, and that the decision-making on EC health programmes has hitherto not been transparent. Recently it has been recognised that it is important, democratically speaking, that NGOs, such as the International Alliance of Patients' Organisations be represented (van der Zeijden, 2000). Indeed, moral issues in medicine have been increasingly seen from the perspective of moral rights.

As there are civil rights, human rights, and consumer rights, rights of minorities and of 'vulnerables', such as the mentally ill, the elderly, the physically handicapped, children, and prisoners – there are indeed patients' rights. The 'rights approach' towards moral issues also takes account of race and gender. It can further be linked to the new ecology and the women's and students' movements of the 1960s, of which the patients' rights movement can be seen as an extension and continuation. Altogether these movements are an expression of deep socio-cultural changes in the Western world. The basic document behind all of them has been, of course, the *Universal Declaration of Human Rights* by the UN General Assembly in 1948. This was followed by the *European Human Rights Convention* signed in 1950 within the framework of the Council of Europe in Rome, to which 48 states have since signed up. The latter is important in that it entailed, in 1952, the establishment of the European Human Rights Court in Strasbourg. This Court, increasingly called upon, developed an unforeseeably dynamic and expansive jurisdiction.

It is noticeable that, in the 1990s, besides the notion of human rights, that of human dignity has effectively emerged, marking yet another cultural shift.

The next section will attempt a typology of the international ethics codes resulting from these scientific and cultural developments. It will focus on the reasons for, and the mechanisms involved in, their genesis by looking more closely at two typical codes from the perspective of an NGO, and an IGO, respectively. This will show some of their distinguishing features.

The WMA Declarations of Helsinki I–VI

One code illustrates the traditional approach at regulating moral issues in medicine. It endorses the paternalistic ethos of the health care professions that foster patients' beneficence as defined by professionals. This 'well-being type' of code can be exemplified by the WMA's *Declaration of Geneva*, its *International Code of Medical Ethics* (see above) and its approach to research on human beings as expressed in the first version (1964) of the *Declaration of Helsinki*. Rather than protecting the freedom and the rights of the patient-subject, as a human rights perspective would suggest, *Helsinki I* was designed to allow a continuation of research on humans. The rights that had been so firmly proclaimed at Nuremberg – and by the WMA *Principles* of 1954 – were 'eroded down to a conditional prerogative of the clinician researcher' (Baker, 1998: 322). This setback in terms of contents was partially corrected in the amended versions (*Helsinki II–V*; 1975 in Tokyo, 1983 in Venice, 1989 in Hong Kong and 1994 in Somerset West). The Tokyo amendment (*Helsinki II*), for instance, introduced the concept of review by an appropriate ethics committee (or Institutional Review Board, IRB) prior to the onset of a project. While these versions *II–V* of the *Declaration* re-endorsed the Nuremberg principles and, by contrast, made it clear that they extended to both research combined with professional care and to non-therapeutic research, they still allowed researchers greater freedom from compulsory consent obligations (Baker, 1998; Winslade and Krause, 1998). While consent could be avoided in certain conditions, those legally or otherwise incapable were especially protected. Furthermore, the personal, non-transferable responsibility of the researcher to assure the ethical quality of the patient-subject's consent, and the right to leave the *experiment* if his/her personal condition seems to demand it (Markers 11 and 12 of Table 5.1), were not incorporated in any of the versions of the *Helsinki Declaration* (Herranz, 1998).

In this context it must be stressed that the WMA as a private organisation represents the interests of the national professional bodies, which it reunites (one per member country only). Its general assembly has so far voted 76 ethics codes, i.e. declarations, statements, recommendations and resolutions, most of whom are hardly known – as for instance the *Declaration of Lisbon on Patients' Rights* (1981, updated 1995; Fluss, 1999). They have no legal force *per se*. But such 'soft law' may become 'hard law' when integrated into legal documents. The WMA has no democratic legitimacy either, for only a part of a nation's doctors is actually represented. The number of national delegates voting at the general assembly depends on the number announced as paying the membership fee to the WMA. This number is arbitrary in that it depends on the financial commitment a national professional organisation is willing to make. The German *Bundesärztekammer*, for instance, pays for some 30 per cent of its members which equals 10 votes in the WMA general assembly, whereas the USA have 12 votes and France 7. The Netherlands did not pay for some time and Switzerland once formally left the WMA and thus both countries temporarily had no vote (Doppelfeld, 2000b). The British Medical Association resigned in the 1970s in protest against the admission of South Africa. It was re-admitted only in 1994. In that year, the membership numbered 64 countries (*Bull. Med. Ethics*, 1994).

It is not clear how and by whom the WMA's ethics codes are formulated (Doppelfeld, 2000a). Certainly they do not follow a consistent philosophical concept of ethics. Rather they reflect compromises arrived at among doctors. An exception was the proposal for a completely new form of the *Helsinki Declaration* presented by the American Medical Association led by Yale philosopher Robert Levine (1997). While the amended versions *II–V* corresponded to some extent to a rebirth of the *Nuremberg Code* from the mid-1970s to the mid-1990s, this proposal for *Helsinki VI* seems to reflect a tendency to deregulation in view of commercial interests in the new biotechnologies (Angell, 2000; Weatherall, 2000; Tröhler 2000b). It generated a controversial debate within the profession and the WMA itself as to whether human research should be related to social benefits at the price of once more reducing the rights of the individual participating (Klinkhammer, 2000). This would be contrary to a number of ethics codes issued by other NGOs and by non-professional IGOs such as the Council of Europe (see below). And such a hierarchy of values, calling to mind eighteenth and nineteenth century attitudes, is certainly not shared throughout the present-day world. A second nearly unaltered proposal by Levine was withdrawn in 1999; the WMA opted for amendments of the latest version (*V*), that of Somerset West (1994).

Finally, in October 2000, the General Assembly at Edinburgh adopted a version in which the fundamental distinction (introduced in the 1964 version) between diagnostic or therapeutic and non-therapeutic biomedical research (i.e., without implying direct value to the person subjected to it) was abolished. In the specific paragraphs allowing medical research to be combined with medical care 'only to the extent that the research is justified by its potential prophylactic, diagnostic or therapeutic value *to the patient*', the latter three words were deleted. This means that social benefit has become uppermost.

In agreement with this outlook, the critical point – formerly listed under 'non-therapeutic research' – stating 'the interest of science and society should never take precedence over considerations related to the well-being of the subjects' – was subtly changed to 'considerations related to the well-being of the human subject should take precedence over the interests of science and society' (WMA, 2000b). This wording represented a compromise compared to the original Levine proposal. This was due to a considerable extent to the intervention of the German Federal Medical Association alarmed by the results of recent scholarly research on medicine during National Socialism.

Indeed, a new form of democracy was practised by having the Levine and subsequent amendment proposals circulated to the national medical associations who then reported to a Workgroup of the WMA (WMA, 2000a). Reflecting the growing connection between industrial interests and clinical research, this new version of *Helsinki (VI)* includes the obligation to adequately inform 'each potential subject of the sources of funding, any possible conflict of interest and institutional affiliations of the researcher' (WMA, 2000b, § 22). These are clear indications of a reorientation of the medical profession to contradictory, yet inevitable, economic constraints.

Of course, the application of the *Declaration* depends on its compatibility with national law and international conventions. In the next section, therefore, examples of ethics codes issued by inter-governmental organisations will be analysed. Their perspective has, either surprisingly or not, changed remarkably as one looks back on the past.

The Council of Europe and its Ethics Codes

The Council of Europe, for instance, has a long-standing history in matters of human rights, particularly in relation to medicine and biology. Its Parliamentary Assembly voted a *Recommendation and Resolution on the Rights of the Sick and Dying* in 1976 followed by one on the *Situation of the Mentally Ill* in 1977, and a further 11 recommendations followed until 1999. The Committee of Ministers voted 21 resolutions, recommendations and protocols thus far (Fluss, 1999). In contrast to the well-being type of code, which addresses scientists' duties, all these represent the 'rights-type' approach to moral issues in medicine. They stress the participants' rights.

As genetic engineering became possible around 1975, the Ministers of Justice of the member states created the *Comité ad hoc des experts en bioéthique* (CAHBI) three years later (1978). Each member state had one vote, but a free number of delegates (usually academics and/or civil servants) depending on a state's political and financial commitment were included. The CAHBI worked many years on a *Convention on Human Artificial Procreation*, which collapsed in the end because of the uncompromising attitude of one member country, Liechtenstein. It was finally published as *Principles* by the CAHBI in 1989. As other problematic ethical fields in medicine were being debated in some of the member states, the Ministers of Justice in 1990 replaced the CAHBI by a standing Steering Committee on Bioethics (*Comité Directeur de Bioéthique*, CDBI). Again, all member states (48 in the year 2000) are represented by one vote, with Canada, the Holy See, the International Federation of Scientific Societies, Israel, and the USA having observer status. The CDBI has also installed task forces working on specific fields such as organ transplants and human genetics. But the main task desired by the ministers was the elaboration of a convention on ethical issues in medicine and biology. The aim was to promote informed consent, considered a patient right, in human research and in daily medical practice. They also wanted to bring order to the array of national and international codes. The committee of ministers soon realised that this was too restrictive an approach for successful political campaigning and insisted on including human rights and human dignity as issues. This led to a change in the title which now reads as the *Convention on the Protection of Human Rights and Dignity of the Human Being with Regard to the Application of Biology and Medicine (HRBM)* or in shortened form *Convention on Human Rights and Biomedicine*. Thus, in terms of contents, the document is a hybrid. A philosophically stringent system could not be implemented.

The *Convention* was drafted according to the conventional methods for arriving at such international agreements. They end up as compromises, sometimes consciously formulated in abstract, even incomprehensible terms – which does not, in the end, serve legal purposes. In the specific case of the *HRBM Convention*, the work of the CDBI was not very efficient, since many of the 60–80 delegates were neither particularly knowledgeable in international law nor on human rights issues and, as usual, the various legal systems had to be taken into account. Many points necessitated tough negotiations and the readiness to compromise (Reusser, 2000). The *Convention* was finally modelled upon that on human rights. Core paragraphs have a commentary to help interpretation. Additional protocols allow a rapid response to new issues. The political nature of this document is evident in the proviso that any country with its own legal regulations for a given issue need not follow the convention in this respect. In consequence some countries introduced laws prior to signing or ratifying the *Convention*.

The core paragraphs concern:

1 the formulation of the *Convention*'s aim, namely the protection of the dignity of the human being 'as soon as life begins' (which is defined nowhere), equal access to health care and the prioritising of the interests of the individual over those of society – an important statement in the context of human experimentation;
2 the requirement of informed consent;
3 the protection of the private sphere;
4 the regulation of the human genome, forbidding discrimination, choice of sex, intervention in the germ line, etc.;
5 the regulation of human research including points 1–3 above;
6 administrative directives, notably for updating the *Convention*, e.g. a complete revision foreseen five years after it takes effect (Council of Europe, 1997).

In 1994 a first draft was ready for national hearings, and in December 1996 the final version was voted through by the Committee of Ministers and presented for signature by governments and separate ratification according to national constitutions. So far, 28 countries have signed and six have ratified the *Convention*, signifying that it is effective and that they must take appropriate legal measures. The national hearings were a new feature. Previous international conventions were drawn up without this democratic instrument. The hearings evidenced serious concerns about research on persons not able to consent and research which is not in their own interest but in that of the population represented.

Summary and Conclusion

Research on human beings gives an example of the slow progress made concerning an ethical need in Western-style medicine, namely that for protection of the participants. Moral issues such as the notion of neglecting safety 'for the sake of experiment' and of 'benefit for the patient' were discussed by some members of the profession as early as the eighteenth century. These issues have, however, often been misconstrued by doctors and the public alike. Many people thought that a therapeutic experiment, lest it constitute an abuse, should guarantee a successful outcome for the well being of every patient. This is not what 'benefit for the patient' means. Rather the benefit of a therapeutic trial lies, indirectly, in the reduction of uncertainty about a beneficial, harmful or non-effective intervention. A directly favourable outcome for an individual patient cannot, therefore, be predicted, otherwise the experiment need not be performed. This fact must not be confounded with the requirement of taking appropriate measures to diminish potential harm in an experiment – 'for the benefit of the patient'.

A legal tradition taking shape around 1900, on a national and, with the *Nuremberg Code* of 1947, on an international level, grew in importance. The *Code* was endorsed in a very few places and lacked positive echo, even in the USA immediately after its inception. The WMA changed the protective attitude of its 1954 *Principles* when issuing the 'paternalistic' *Declaration of Helsinki (I)* in 1964 in reaction to serious abuses, such as the thalidomide case. This allowed research to proceed and prevented legislative curtailment. Ten years later, with the amended *Helsinki Declaration (II)*, patient protection was improved, to be qualified again twenty years later, towards the end of the twentieth century, in the updated *Helsinki Declaration (VI)*. This is in contrast to the 'human rights and human dignity' approach of IGO codes, most particularly of the *Convention of Human Rights and Biomedicine* of the Council of Europe. However, new research scandals in the 1990s and the possibility of exporting research to developing countries (Lurie and Wolfe, 1997), have proved the irrelevance of the WMA's and many other international 'soft law' ethics codes in relation to international law. Beyond their legal status, the influence of both codes and laws was defined by their relevance to contemporary thought, whether they addressed conflicts perceived by society and whether they offered any means of solving them.

Incontestably, however, the 'soft law' codes offer some positive features compared to law and international conventions, namely specificity and flexibility. Their legitimacy, in contrast to democratically passed laws, is, however, problematic (Matthieu, 1998), as the example of the WMA codes has shown. Therefore, the process of how these codes are conceived is important. It should be made transparent, which, as it seems, has not always been the case for the WMA codes. The issue of international regulations raises the question of different ethical views and cultures, not only regarding the universality of conflicts, but also the means of solving them, for instance, via informed consent and other principles which need not necessarily be important to everyone.

If the purpose of an NGO-code or an IGO-document is to contribute to legal liability, their success in judging the validity of human research in view of continuous abuses over the past 50 years must be pondered. The future looks more optimistic for codes fostering patient rights and perhaps human dignity in daily medical practice (Reusser, 2000).

But ethics codes can also have other purposes, namely the articulation of idealistic aspirations. They can play the role of emblematic symbols (Smith, 1996; Leven, 1997), 'express fundamental values, such as respect for persons and scientific integrity ... set standards for moral criticism of medical practices or policies that may violate the rights of human subjects' (Winslade and Krause, 1998: 140). Their general nature has, in certain cases, actually 'dramatized the need for national and international documents with binding authority' (Perley et al., 1992: 160). Of course, they can also serve as moral cover-ups. Last, but not least, codes may serve as teaching aids. With legislation intervening in an ever more detailed way in all matters of medicine, science and human rights, it is important that codes are being observed, whatever their legal status, in order that 'Hippocrates' not *hypocrisia* reign in human research.

References

Angell, M. (2000), 'Is Academic Medicine for Sale?', *New England Journal of Medicine*, vol. 342, pp. 1516–18.

Annas, G.J. and Grodin M.A. (eds) (1992), *The Nazi Doctors and the Nuremberg Code. Human Rights in Human Experimentation*, Oxford University Press, New York and Oxford, pp. 149–73.

Arnold, P. and Sprumont, D. (1998), 'The "Nuremberg Code": Rules of Public International Law', in Tröhler, U. and Reiter-Theil, S. (eds), *Ethics Codes in Medicine: Foundations and Achievements of Codification since 1947*, Ashgate, Aldershot, pp. 84–96.

Baker, R. (1993), 'Deciphering Percival's Code', in Baker R., Porter R. and Porter D. (eds), *The Codification of Medical Morality*, vol. 1: *Medical Ethics and Etiquette in the Eighteenth Century*, Kluwer, Dordrecht, Boston, London, pp. 179–211.

Baker, R. (1998), 'Transcultural Medical Ethics and Human Rights', in Tröhler, U. and Reiter-Theil, S. (eds), *Ethics Codes in Medicine: Foundations and Achievements of Codification since 1947*, Ashgate, Aldershot, pp. 312–31.

Bassiouni, M.C., Baffes, T.G., Evard, J.T. (1981), 'An Appraisal of Human Experimentation in International Law and Practice: The Need for International Regulation of Human Experimentation', *Journal of Criminal Law and Criminology*, vol. 72, pp. 1597–666.

Bernard, C. (1949), *Introduction à la médecine experimentale (1865)*, publ. as *An Introduction to the Study of Experimental Medicine*, trans. by H.C. Greene, Schuman, New York.

Blom, K. (1973), 'Armauer Hansen and Human Leprosy Transmission. Medical Ethics and Legal Rights', *International Journal of Leprosy*, vol. 41, pp. 199–207.

Brennan, T.A. (1999), 'Proposed Revisions to the Declaration of Helsinki – Will they Weaken the Ethical Principles Underlying Human Research?', *New England Journal of Medicine*, vol. 341, pp. 527–30.

Bulletin of Medical Ethics (1994), 'World Medical Association is Reanimated' (Editorial), *Bulletin of Medical Ethics*, no. 101, September 1994, pp. 3–4.

Bynum, W.F. (1988), 'Reflections on the History of Human Experimentation', in Spicker, S.F. et al. (eds), *The Use of Human Beings in Research*, Kluwer, Dordrecht, Boston, London, pp. 29–46.

Council of Europe (1997), 'Convention for the Protection of Human Rights and Dignity of the Human Being with Regard to the Application of Biology and Medicine': *Convention on Human Rights and Biomedicine*, Orviedo, U. IV, 1997, European Treaty Series, no. 164.

Currie, J. (1804), *Medical Reports on the Effects of Water, Cold and Warm as a Remedy in Fever and Febrile Diseases*, vol. 2, McCreery and Cadell, Liverpool and London.

Deutsch, E. (1997), 'The Nuremberg Code: The Proceedings in the Medical Case, the Ten Principles of Nuremberg and the Lasting Effect of the Nuremberg Code', in Tröhler, U. and Reiter-Theil, S. (eds), *Ethics Codes in Medicine: Foundations and Achievements of Codification since 1947*, Ashgate, Aldershot, Brookfield, Singapore, Sydney, pp. 71–83.

Doppelfeld, E. (1999), 'Generalversammlung des Weltärztebundes: Offene Fragen, ungelöste Probleme', *Deutsches Ärzteblatt*, vol. 96, pp. C–2297–9.

Doppelfeld, E. (2000a), 'Weltärztebund – Probe für die Glaubwürdigkeit', *Deutsches Ärzteblatt*, vol. 97, pp. A–1587–92.

Doppelfeld, E. (2000b), Paper read at the international symposium 'Das Menschrechtsübereinkommen des Europarates – taugliches Vorbild für eine weltweit geltende Regelung?', Heidelberg, Akademie der Wissenschaften, 19 September 2000, and personal communication.

Drinan, R.F. (1992), 'The Nuremberg Principles in International Law', in Annas, G.J. and Grodin, M.A. (eds), *The Nazi Doctors and the Nuremberg Code. Human Rights in Human Experimentation*, Oxford University Press, New York and Oxford, pp. 174–82.

Elkeles, B. (1996), *Der moralische Diskus über das medizinische Menschenexperiment im 19. Jahrhundert*, Gustav Fischer, Stuttgart, Jena, New York.

Fluss, S.S. (1999), 'International Guidelines on Bioethics', *EFGCP News*, December 1999, supplement.

Geison, G.L. (1995), *The Private Science of Louis Pasteur*, Princeton University Press, Princeton, NJ.

Glantz, L.H. (1992), 'The Influence of the Nuremberg Code on U.S. Statutes and Regulations', in Annas, G.J. and Grodin, M.A. (eds), *The Nazi Doctors and the Nuremberg Code. Human Rights in Human Experimentation*, Oxford University Press, New York and Oxford, pp. 183–200.

Gradmann, C. (2001), 'Robert Koch and the Pressures of Scientific Research: Tuberculosis and Tuberculin', *Medical History*, vol. 45, pp. 1–32.

Guthrie, G.J. (1815), *On Gun-shot Wounds of the Extremities*, Longman, London.

Herranz, G. (1998), 'The Inclusion of the Ten Principles of Nuremberg in Professional Codes of Ethics: An International Comparison', in Tröhler, U. and Reiter-Theil, S. (eds), *Ethics Codes in Medicine: Foundations and Achievements of Codification since 1947*, Ashgate, Aldershot, Brookfield, Singapore, Sydney, pp. 127–39.

Howard-Jones, N. (1982), 'Human Experimentation in Historical and Ethical Perspectives', *Social Science and Medicine*, vol.16, pp. 1429–48.

Klinkhammer, G. (2000), 'Medizinische Forschung am Menschen. Abkehr von einheitlichen Standards', *Deutsches Ärzteblatt*, vol. 97, pp. A–2205f.

Lederer, S. (1995), *Subjected to Science. Human Experimentation before the Second World War*, Johns Hopkins University Press, Baltimore and London.

Leven, K.-H. (1997), 'Der hippokratische Eid im 20. Jahrhundert', in Toellner, R. and Wiesing, U. (eds), *Geschichte und Ethik in der Medizin*, Fischer, Jena, pp. 111–29.
Levine, R.J. (1999), 'The Need to Revise the Declaration of Helsinki', *New England Journal of Medicine*, vol. 341, pp. 531–4.
Lurie, P. and Wolfe, S.M. (1997), 'Unethical Trials of Interventions to Reduce Perinatal Transmission of Human Immmunodeficiency Virus in Developing Countries', *New England Journal of Medicine*, vol. 337, pp. 853–6.
Macbride, D. (1764), *Experimental Essays*, Millar, London.
Maehle, A.-H. (1999a), *Drugs on Trial: Experimental Pharmacology and Therapeutic Innovation in the Eighteenth Century*, Rodopi, Amsterdam and Atlanta, GA.
Maehle, A.-H. (1999b), 'Professional Ethics and Discipline: The Prussian Medical Courts of Honour, 1899-1920', *Medizinhistorisches Journal*, vol. 34, pp. 309–38.
Maehle, A.-H. (2000), 'Assault and Battery, or Legitimate Treatment? German Legal Debates on the Status of Medical Interventions without Consent, c. 1890–1914', *Gesnerus*, vol. 57, pp. 206–21.
Maehle, A.-H. and Tröhler, U. (1990), 'Animal Experimentation from Antiquity to the End of the Eighteenth Century: Attitudes and Arguments', in Rupke, N.A. (ed.), *Vivisection in Historical Perspective*, Routledge, London and New York, pp. 14–47.
Maio, G. (2001), 'Ärztliche Ethik als Politikum. Zur französischen Diskussion um das Humanexperiment nach 1945', *Medizinhistorisches Journal*, vol. 36, pp. 35–80.
Mathieu, B. (1998), 'Ethical "Norms" and the Law: Legitimacy of "Experts" or Democratic Legitimacy', in Tröhler, U. and Reiter-Theil, S. (eds), *Ethics Codes in Medicine: Foundations and Achievements of Codification since 1947*, Ashgate, Aldershot, Brookfield, Singapore, Sydney, pp. 163–84.
McCullough, L.B. (1998), *John Gregory's Writings on Medical Ethics and Philosophy of Medicine*, Kluwer, Dordrecht, Boston, London.
McLean, C. (1817–18), *Results of an Investigation Respecting Epidemic and Pestilential Diseases*, Underwood, London.
Medical Research Council (1953), 'Draft Statement (revised) on Clinical Investigations', 9 October 1953, signed by H. F. Hinsworth, Manuscript MRC.53/518/B.
Minister der Geistlichen, Unterrichts- und Medizinal-Angelegenheiten (1901), 'Anweisung an die Vorsteher der Kliniken, Polikliniken und sonstigen Krankenanstalten' (dated 29 December 1900), *Centralblatt für die gesamte Unterrichts-Verwaltung in Preußen*, 1901, pp. 188–9.
Moll, A. (1902), *Ärztliche Ethik. Die Pflichten des Arztes in allen Beziehungen seiner Thätigkeit*, Enke, Stuttgart.
Perley, S. et al. (1992), 'The Nuremberg Code: An International Overview', in Annas, G. and Grodin, M.A. (eds), *The Nazi Doctors and the Nuremberg Code. Human Rights in Human Experimentation*, Oxford University Press, New York and Oxford, pp. 149–73.
Reich, W.T. (ed.) (1995), *Encyclopedia of Bioethics*, 5 vols, Simon and Schuster Macmillan, New York.
Reusser, R. (2000), Paper read at the international symposium 'Das Menschrechtsübereinkommen des Europarates – taugliches Vorbild für eine weltweit geltende Regelung?', Heidelberg, Akademie der Wissenschaften, 19 September 2000, and personal communication.
Rothman, D. (1995), 'Research, Human: Historical Aspects', in Reich, W. T. (ed.), *Encyclopedia of Bioethics*, Simon and Schuster Macmillan, New York, vol. 4, pp. 2238–58.
Rothman, D. (1998), 'The Nuremberg Code in Light of Previous Principles and Practices in Human Experimentation', in Tröhler, U. and Reiter-Theil, S. (eds), *Ethics Codes in Medicine: Foundations and Achievements of Codification since 1947*, Ashgate, Aldershot, Brookfield, Singapore, Sydney, pp. 50–59.

SAMW (Swiss Academy of Medical Sciences) (2001), *Ethische Richtlinien – Ethics Guidelines and Recommendations*, www.samw.ch.

Sass, H.-M. (1988), 'Comparative Models and Goals for the Regulation of Human Research', in Spicker, S.F. et al. (eds), *The Use of Human Beings in Research*, Kluwer, Dordrecht, Boston, London, pp. 47–89.

Sauerteig, L. (2000), 'Ethische Richtlinien, Patientenrechte und ärztliches Verhalten bei der Arzneimittelerprobung (1892–1931)', *Medizinhistorisches Journal*, vol. 35, pp. 301–32.

Schöne-Seifert, B. (1999), 'Defining Death in Germany', in Youngner, S.J., Arnold, R.M. and Schapiro, R. (eds), *The Definition of Death, Contemporary Controversies*, Johns Hopkins University Press, Baltimore and London, pp. 257–71.

Schuster, E. (1997), 'Fifty Years Later: The Significance of the Nuremberg Code', *New England Journal of Medicine*, vol. 337, pp. 1436–40.

Smith, D.C. (1996), 'The Hippocratic Oath and Modern Medicine', *Journal of the History of Medicine and Allied Sciences*, vol. 51, pp. 484–500.

Smith, R.G. (1994), *Medical Discipline: The Professional Conduct Jurisdiction of the General Medical Council, 1858–1990*, Clarendon Press, Oxford.

Spicer, C.M. (1995), 'Nature and Role of Codes, and Other Ethics Directives', in Reich, W.T. (ed.), *Encyclopedia of Bioethics*, Simon and Schuster Macmillan, New York, vol. 5, pp. 2605–12.

Starr, C. (1998), *Blood. An Epic History of Medicine and Commerce*, New York, A. Knopf.

Tröhler, U. (1993), 'Surgery (modern)', in Bynum, W.F. and Porter, R. (eds), *Companion Encyclopedia of the History of Medicine*, Routledge, London and New York, vol. 2, pp. 984–1028.

Tröhler, U. (1998), 'From Rehn's Risky Cardiac Suture (1896) to Routine Cardiac Transplantation (1996): Historical and Ethical Perspectives', *Journal of Cardiovascular Surgery*, vol. 39, supplement 1 to no. 2, pp. 7–22.

Tröhler, U. (1999), 'Das ärztliche Ethos und die Kodifizierung von Ethik in der Medizin', in Bondolfi, A. and Müller, H. (eds), *Medizinische Ethik im ärztlichen Alltag*, EMH, Schweiz. Ärzteverlag, Basle and Berne, pp. 39–61.

Tröhler, U. (2000a), *To Improve the Evidence of Medicine: The Eighteenth Century British Origins of a Critical Approach*, Royal College of Physicians, Edinburgh.

Tröhler, U. (2000b), 'Asilomar-Konferenz zu Sicherheit in der Molekularbiologie von 1975: Rückschau und Ausblick', *Schweiz. Aerztezeitung*, vol. 81, pp. 1585–7.

Tröhler, U. and Maehle, A.-H. (1990), 'Anti-Vivisection in Nineteenth-Century Germany and Switzerland: Motives and Methods', in Rupke, N.A. (ed.), *Vivisection in Historical Perspective*, Routledge, London and New York, pp. 149–87.

Übelhack, B. (2002), *Ärztliche Ethik – Eine Frage der Ehre? Die Prozesse und Urteile der ärztlichen Ehrengerichtshöfe in Preussen und Sachsen 1918–33*, Lang, Frankfurt/M., in press.

van der Zeijden, A. (2000), 'Citizens and Patients as Partners in Decision-Making and Implementation', *Issues in Eureopean Health Policy*, vol. 3, p. 8.

Veatch, R.M. (1995), 'Medical Codes and Oaths II: Ethical Analysis', in Reich, W.T. (ed.), *Encyclopedia of Bioethics*, Simon and Schuster Macmillan, New York, vol. 3, pp. 1427–35.

Vollmann, J. and Winau, R. (1996), 'Informed Consent in Human Experimentation before the Nuremberg Code', *British Medical Journal*, vol. 313, pp. 1445–7.

Weatherall, D. (2000), 'Academia and Industry: Increasingly Uneasy Bedfellows', *Lancet*, vol. 355, pp. 1574f.

WMA (World Medical Association, Inc.) (2000a), *Documents WG/DoH*, July 2000; 17.C/WW3/2000.

WMA (World Medical Association, Inc.) (2000b), 'The Declaration of Helsinki' (revised version adopted 3 October 2000 in Edinburgh), *Bulletin of Medical Ethics*, October 2000, pp. 8–11.

Winslade, W.J. and Krause, T.L. (1998), 'The Nuremberg Code Turns Fifty', in Tröhler, U. and Reiter-Theil, S. (eds), *Ethics Codes in Medicine: Foundations and Achievements of Codification since 1947*, Ashgate, Aldershot, pp. 140–62.

Chapter 6

Ethical Aspects of Life-Saving and Life-Sustaining Technologies

Bryan Jennett

Introduction

This chapter deals from a doctor's perspective with the same dilemma as Chapter 7. That dilemma arises from the technological developments in recent years that now enable doctors to save and sustain the lives of many patients who would previously have died. Some rescued patients lead lives of good enough quality for long enough after such interventions that no one doubts that medical treatment was of benefit. However, concern is increasingly voiced that these technologies sometimes do no more than prolong the process of dying, or extend lives of very poor quality. Doctors are then often regarded as having done more harm than good. These doubts about medical care have emerged not only because, with such technologies increasingly available, doctors have themselves become concerned about delivering non-beneficial treatment. They reflect also a public who are better informed about medical treatments and approach them from a consumerist standpoint. As a result patients and their families increasingly challenge the traditional, paternalistic role of doctors in making decisions that profoundly affect the lives of their patients.

Perception of a Problem

In 'An Appeal to Doctors' in the *Lancet* over 30 years ago a Cambridge Professor of Physics referred to 'the hardship caused to patients and relatives by the unrelenting application of modern techniques'(Thompson, 1969). 'Matters might improve,' he suggested, 'if medical ethics were to make the useless prolonging of life an offence.' The next year an anonymous physician in the *Lancet* asked, 'I wonder whether our profession is fully aware of the deep and widespread concern in the general public at not being allowed to die' (FRCP, 1970). These comments received a cool reception in an editorial, which expressed fear about any decision that might be considered to have led to a patient dying sooner than he might otherwise have done (*Lancet*, 1970).

In contrast to this the approach to what are now often referred to as 'treatment-limiting decisions at the end of life' has become increasingly formalised in the USA. By 1968, when intensive care units were becoming widely established, doctors in them began calling for 'death with dignity' and 'natural death' as contrasted with death on a ventilator with tubes and wires from several body orifices. Recognition that doctors might sometimes need help with decision-making in this difficult area led in the USA to what was sometimes termed 'the flowering of bioethics', with more than 70 Professors of Medical Ethics appointed and many hospitals employing ethicists to advise professionals and administrators. Ethics Committees comprising doctors, nurses, moral philosophers, lawyers and others were established in hospitals to debate the principles of such decision-making and, when requested, to discuss with families and doctors decisions about individual patients that had become contentious.

Ethical Aspects of Life-Prolonging Treatment

Philosophers and lawyers now frequently challenge doctors to reconsider their duties to, and the rights of, their patients in the light of the importance of self-determination, as expressed in the ethical principle of patient autonomy and the legal principle of consent. Doctors are urged to give due weight to patient preferences when deciding about treatment, but to consider also the three other principles of medical ethics – beneficence, non-maleficence and social justice in the use of health care resources. Inevitably in practice there may be conflicts between these four principles, in particular when the doctor and the patient are each judging the balance between the probable benefit and harm associated with treatment options. But there may also be moral conflicts. One is between prolonging life on the one hand and relieving suffering and minimising indignity on the other. The sanctity of life principle is not absolute, and only a minority of vitalists maintain that life *per se* is always good, no matter what the state of the patient or his wishes. Another conflict may be between the rights of patients and the duties of doctors. Patients have a right to information, autonomy and respect, whilst doctors have duties not only to individual patients and their families but also to other carers, to competing patients, to their profession, and to society as reflected in the law. Gone are the days when a doctor simply did everything possible for every patient for as long as possible, and no questions were asked.

Some Solutions in America

In 1976 a group of intensivists published a paper on 'Optimum care for the hopelessly ill', suggesting that for some patients the less doctors did the better

(Critical Care Committee of the Massachusetts General Hospital, 1976). The same journal issue had papers also on do-not-resuscitate orders, to protect patients already terminally ill from the indignity of cardio-pulmonary resuscitation (CPR) with defibrillators and ventilators (Rabkin et al., 1976); and on living wills or advance directives, to enable patients while competent to indicate what treatments they would not want if critically ill and unable themselves then to refuse treatment (Bok, 1976). That same year a New Jersey court granted the request of the Roman Catholic father of the comatose teenager Karen Quinlan to have her ventilator discontinued so that she might be allowed to die, a decision supported by the local bishops. In the next few years there were more than 100 decisions in US courts allowing treatment withdrawal from competent and incompetent patients. In 1983 a President's Commission, that included doctors, nurses, philosophers and jurists, recommended that treatment withdrawal in certain circumstances was legally and morally acceptable and was considered good medical practice. This specifically included withdrawing artificial nutrition and hydration from permanently vegetative patients. The American Medical Association (AMA) endorsed this position, and in the late 1980s several American judges declared that it was unnecessary and indeed inappropriate to bring such cases to court unless there was some specific dispute between involved parties. Otherwise the decision should be a matter for discussion and agreement between doctors and patients or those close to them. There followed requirements that hospitals and nursing homes have formal protocols for treatment withdrawal. Later, after the US Supreme Court decided in favour of the withdrawal of tube feeding from Nancy Cruzan, a vegetative patient, came renewed calls for more people to make advance directives and also the Patient Self-determination Act. This requires all patients on admission to hospital to be told about their right to refuse treatment, to make an advance directive or to appoint a surrogate decision-maker in the event of their becoming incompetent in the future. In spite of all these developments an editorialist in the *New England Journal of Medicine* in 1990 stated: 'The very high suicide rate in older Americans is due partly to concern that they may be unable to stop treatment if hospitalised. Some people now fear living more than dying, because they dread becoming prisoners of technology' (Angell, 1990).

Situation in Britain

Medical comment in Britain initially suggested that this was largely an American problem, the product of a supposed tendency to over-treatment resulting from too readily available medical technology in a litigious society. This view maintained that the formal and legalistic responses developed in the USA were unnecessary in Britain. To quote one eminent physician: 'In Britain the medical team is under few restraints, from the family, ethics and least of all the law, to act as it sees fit in the best interest of the patient' (Bayliss, 1982). Two decades later this assertion that 'doctors rule OK' sounds unacceptably paternalistic. In 1988 the British Medical Association (BMA) in its Euthanasia Report, while completely rejecting any move towards euthanasia, made clear that treatment withdrawal from some hopelessly ill patients was appropriate, including tube feeding from vegetative patients (British Medical Association, 1988). The following year an editorial subtitled 'Learning from America' suggested that the adoption of protocols for limiting life-sustaining treatment might be beneficial (Williams, 1989). In 1993 in the High Court, Appeal Court and the House of Lords nine judges unanimously declared as legal the withdrawal of tube feeding from Tony Bland, a teenage vegetative patient. Their comments in those courts led to the issuing of guidance that requires formal court approval for every such withdrawal (Official Solicitor, 1996). More than 25 vegetative cases have now been heard in the High Court and all have been approved. In response to the Bland case the BMA produced guidelines for the withdrawal of tube feeding from vegetative patients (British Medical Association, 1993), and these were later expanded by the Royal College of Physicians of London (1996). Most recently the BMA has issued guidance on limiting treatment, not only of vegetative patients but of others who are hopelessly ill (British Medical Association, 1999, 2001).

Some Practical Clinical Issues

Life-prolonging treatments that may be limited include those that save life such as CPR, emergency surgery and drugs that deal with infections, irregularities of the heart beat or low blood pressure and also treatments that sustain life, such as the technological package of intensive care and substitutes for organ failure, such as mechanical ventilation, dialysis and tube feeding. Patients considered for limited treatment can be classified into three groups, for each of which the reasons for such a decision are different. There are those who, being previously well, have a sudden unexpected life-threatening crisis; this might be a heart attack, brain damage from head injury or a stroke, from lack of oxygen due to choking or near drowning or a medical accident, often under anaesthesia. Doctors need time (usually a few days) to assess the response of such patients to emergency treatment before a decision can be made that the outlook is hopeless, based on the failure of a trial of treatment. Quite different are patients with progressive disorders in whom a predictable crisis or relapse occurs, and for whom a prior decision can often be made, after consultation with the patient or family, to withhold a specific intervention when such an event occurs. These include patients with progressive failure of various organs – heart, lungs, brain, kidneys or liver, those with extensive paralysis, or advanced cancer and those in the late stages of AIDS. Different again are patients with slowly progressive or static conditions associated with very poor quality of life, already dependent on life-sustaining treatment, for whom the question of treatment withdrawal may arise without the intervention of any new medical crisis.

The withdrawal of life-sustaining treatment from patients who might otherwise live for years is sometimes seen as more problematic than limiting life-prolonging measures in situations when death is expected in days or weeks even with all possible treatment. However, such action is best regarded as an extension of the good practice of withdrawing treatment once a trial, albeit over an extended period, has shown there to be no benefit in that there is no hope of recovery from an unacceptable quality of life. Many such cases have been decided in the courts in the USA and some in Britain both for incompetent patients and also for competent patients with severe neurological disorders who request withdrawal of ventilation or feeding or ask for dialysis to be discontinued. The cause of death in such cases is considered to be the antecedent medical condition, the fatal outcome of which has been temporarily delayed by a trial of treatment.

Treatment may be considered hopeless for one of three distinct reasons. One is when treatment would be futile because it would be ineffective in saving or prolonging life. More commonly an acute intervention might postpone death but for such a brief period, or for longer with such poor quality of life, that the burden of treatment is considered disproportionate to the benefit. The third situation is when a patient already has such poor quality of life that continued survival is not considered a benefit. Decisions to limit treatment in one or other of these circumstances are now quite frequently made.

One American estimate was that patients' families or their physicians had in some way planned 70 per cent of deaths occurring in hospital. In a study of intensive care units in San Francisco half the deaths followed a decision to limit treatment, with withdrawal four times more frequent than withholding (Smedira et al., 1990). Of more than 9,000 deaths in the Netherlands 50 per cent of non-acute deaths were hastened by a medical decision, half of them due to treatment limitation, half from the side-effects of drugs to relieve symptoms. That some drugs given to alleviate symptoms have a possible side effect of hastening death is considered ethically acceptable by the (originally Catholic) doctrine of double effect. A study of deaths in surgical wards in Scotland found that some 40 per cent followed a decision to limit treatment. A confidential enquiry into deaths soon after surgery in Britain showed that some 10 per cent of the operations were assessed as having been unnecessary and unjustified by reason of the patient having been already moribund because of the advanced stage of the disease or the patient being very old, often both.

Why is Non-Beneficial Treatment Given?

The reason most often given why doctors sometimes initiate, or persist with, inappropriate treatment is uncertainty about whether the clinical condition is in fact hopeless. Associated with this is the assumption that if there is any possibility of any improvement, however small, the patient must be given the supposed benefit of that doubt. There is, however, reason to question the degree of certainty needed for a decision to limit treatment. A group of neurosurgeons from several countries was asked 'at what level of probability of a poor outcome (death or survival in the vegetative or a very severely disabled state) would a decision to withhold surgery or ventilation after recent severe head injury be justified?' (Barlow and Teasdale, 1986). Most wanted 95 per cent or more certainty of a poor outcome before making such a decision. However, when they were then asked to imagine themselves as the injured patient, many of these surgeons wanted treatment limited at a much lower probability of a poor outcome. In other words, they were unwilling to accept for themselves the risk of brain damaged survival that they would impose on their patients. When 500 Americans were asked their preferences for non-treatment under various circumstances 90 per cent did not want resuscitation, ventilation, emergency surgery or tube feeding if they were vegetative or demented (Emanuel et al., 1991). Moreover, if they were in coma with a small chance of full recovery, more than 50 per cent did not want any such life-saving or life-sustaining measures. These studies show that people do not want life at any price, are fearful of severely disabled survival, and that they recognise that such decisions depend on probability rather than on certainty.

Uncertainty about prognosis may, however, be a valid reason for initiating life-saving treatment when faced with an emergency in a patient previously not seriously ill, because it is only after a trial of treatment has proved unsuccessful that further treatment can be declared to be futile. However, doctors may then persist because they find it more difficult emotionally to withdraw than to withhold treatment, although both lawyers and philosophers have repeatedly asserted that there is no ethical or legal difference between withholding and withdrawing treatment. Unless the logical fallacy of this supposed distinction is recognised not only may patients have long periods of futile treatment but a more sinister consequence is possible. This is that doctors may sometimes be reluctant to initiate treatment that could save a life with good recovery because of fear that if it proves unsuccessful they would be committed to a prolonged period of futile treatment.

Sometimes treatment that is known to be futile is undertaken because of the unwillingness of the doctor to admit to the patient or the family that the situation is hopeless. He may even deny this himself, by calling on prognostic uncertainty as the supposed reason even when all the evidence points clearly to a hopeless situation. This is, however, becoming less common as the ethical imperative to tell patients the truth, in order to allow them to exercise their autonomy, makes it no longer acceptable to deceive patients about their prognosis. It is now commonplace, for example, to reach a stage when active treatment of cancer is discontinued with the patient's agreement. This has been greatly facilitated by the emergence of the hospice movement which ensures that such patients continue to receive a high level of palliative care. Indeed it should always be part of a treatment-limiting decision that all measures that ensure the comfort and dignity of the dying patient should be vigorously applied.

The Question of Consent

Like patients with cancer, those with cardio-respiratory or renal failure and with AIDS are mostly competent and can indicate their wish to have no further treatment. The law is clear that the doctor must comply with such a preference, which is in effect a refusal of consent for continued treatment. Many courts in the United States and the United Kingdom have explicitly stated this. These lawyers emphasise that the doctor must comply with the patient's wish to limit treatment even if he thinks this is unwise, provided the doctor is satisfied that the patient is fully informed, legally competent, and does not suffer from treatable depression. However, many hopelessly ill patients are not competent to make a decision about their treatment. This includes infants, and adults with brain failure, either acute due to intracranial trauma, haemorrhage or infection or to systemic disease affecting cerebral function; or chronic due to mental retardation, dementia or the vegetative state.

The question is how to ensure that incompetent patients can have their autonomy respected and do not have treatment that they would have been likely to refuse had they been competent. An extension of the autonomy of incompetent adults can best be achieved if they have made an advance directive (or living will) or have appointed a proxy decision-maker, to ensure that their preferences about treatment at the end of life will be known and can be respected. Lawyers are clear that a doctor is at no risk from civil or criminal liability for the consequences of complying with a patient's wish not to be treated, whether expressed now or previously. Also the patient's death should not be regarded as resulting from suicide or euthanasia.

However, it is only occasionally that an advance directive is available when making these decisions. Without this the family may be asked to say what they think this patient would have wanted, taking account of the value system of the patient and any comments that might have been made previously about survival in certain states. This is the so-called substituted judgement of the American courts. The English courts prefer that doctors, after consulting families, should decide what is in the best interests of the patient. Again account should be taken of any informal indications the patient might have made in the past about their wishes, and also of what most people in society would want in these circumstances. There is, for example, a widespread consensus that prolonged survival in the vegetative state is not a benefit. Although some consider that the sanctity of life principle should always prevail, many moral philosophers have explicitly rejected this minority view. So also have the several courts that considered the Bland case in England as well as many courts in the USA, and others in Canada, New Zealand, South Africa, Scotland and the Republic of Ireland (Jennett, 2002).

Formalities to Observe when Limiting Treatment

The question of what consultations and formalities should be required of doctors making decisions to limit treatment is best discussed in relation to three different situations. The first is when treatment is considered to be futile because it could not produce any physiological benefit. As doctors are under no moral or legal obligation to provide futile treatment, nor can patients or families demand it, it could be considered unkind and unnecessary to discuss the option of futile treatment with the patient or the family. However, it is probably wise to mention that certain options are being rejected because they can be of no benefit. When the issue is the proportionality of burden and benefit of treatment that offers only a brief or partial respite, some discussion is obviously necessary. This applies both to the initiation or continuation of treatment in a current crisis and to prior discussion of what might be done when a predictable future crisis occurs. Many of these clinical situations are relatively stereotyped, and it is practical and useful to have guidelines to indicate those circumstances under which such a decision might be considered. Devising guidelines provides an opportunity for doctors and nurses together to discuss the principles involved, outside the emotional context of making a decision about a particular patient. This provides the best opportunity for allowing different views to be considered and a consensus reached. The existence of such agreed guidelines then makes it clear that such decisions are accepted as good practice in certain prescribed situations, and to some extent they may make future decisions about individual patients more of a shared responsibility rather than the decision of a single clinician.

The BMA has suggested a formal approach by the medical profession to such decisions, involving a second opinion and the keeping of records of all such cases for later review and audit (British Medical Association, 1999, 2001). Another issue is whether such a decision needs to come to court. Although the American bench has indicated that judicial review of such cases is no longer necessary, the English courts deciding Bland held that withdrawal of tube feeding from future vegetative patients should proceed only after a second medical opinion and a decision in court. However, they indicated that once good practice had been established and the public reassured, guidelines might emerge that would allow such decisions to be reached by a formal process that need not involve court proceedings, and that is the recommendation of the Law Commissions in both England and Wales and in Scotland (Jennett, 2002).

Controversial Issues

It is a matter of judgement whether a specific treatment is of no benefit or of too little benefit to justify it. The competent patient is the best judge of this, and provided he is fully informed his view is not to be challenged. With an incompetent patient without an advance directive the problem arises of judging the quality of life of another person, particularly when the judge is a healthy person. Those who care for seriously handicapped patients often attest to their adaptation to a limited life, and to regarding this as worthwhile. Some use this to argue against the validity of advance directives, on the grounds that the person in a severely disabled state might regard this differently than when that person contemplated it whilst healthy. Another debate is whether tube feeding is medical treatment that can be withdrawn by a doctor or should be regarded as basic care that should never be stopped. The latter view holds that its withdrawal might indicate abandonment of the patient – but all withdrawal protocols emphasise the need to continue a high standard of nursing care. Almost all those bodies in various countries that have considered this formally have agreed that tube feeding is medical treatment that substitutes for failed swallowing in the same way as a ventilator provides artificial breathing and dialysis an artificial kidney. Another issue is the status of the surrogate decision-maker, who in the United States has legal power to make a decision about treatment. That is not so in the UK where it is only a doctor who can decide to limit treatment, although consultation with the family or friends of an incompetent patient is required. The main purpose of such consultation is to attempt to discover what this patient's likely wishes would be, were he able to express his own opinion, in order to respect his autonomy.

Conclusion

There is clearly need for doctors, lawyers and moral philosophers to respond constructively to public concern about inappropriate treatment for hopelessly ill patients. The aim should be to promote compassionate care and to respect the wishes of competent patients. For incompetent patients any previously expressed wishes should be respected, but in the absence of these it is their best interests that should be the focus of concern. In judging these due attention should be paid to the patient's family or friends, but the decision should also reflect the consensus of opinion about such dilemmas in the public at large – all of whom are potentially hopelessly ill patients. They need to be reassured that doctors have thoughtfully considered the many issues that should inform such decisions, and will weigh these carefully in regard to every individual patient.

References

Angell, M. (1990), 'Prisoners of Technology', *New England Journal of Medicine*, vol. 322, pp. 1226–8.

Anon (Editorial) (1970), 'Not Strive Officiously', *Lancet*, vol. 2, p. 915.

Anon (FRCP) (1970), 'Right to Die', *Lancet*, vol. 2, p. 926.

Barlow, P. and Teasdale, G. (1986), 'Prediction of Outcome and the Management of Severe Head Injury: the Attitudes of Neurosurgeons', *Neurosurgery*, vol. 19, pp. 989–91.

Bayliss, R.I.S. (1982), 'Thou Shalt Not Strive Officiously', *British Medical Journal*, vol. 285, pp. 1373–5.

Bok, S. (1976), 'Personal Directions for Care at the End of Life', *New England Journal of Medicine*, vol. 295, pp. 367–9.

British Medical Association (1988), *Euthanasia*, BMA, London.

British Medical Association (1993), 'Medical Ethics: Persistent Vegetative State', *BMA Supplementary Annual Report*, BMA, London, pp. 2–4.

British Medical Association (1999, 2001), *Withholding and Withdrawing Life-Prolonging Medical Treatment: Guidance for Decision-Making*, BMJ Books, London; 2nd edition 2001.

Committee for Critical Care of the Massachusetts General Hospital (1976), 'Report on Optimum Care for the Hopelessly Ill', *New England Journal of Medicine*, vol. 295, pp. 363–4.

Emanuel, L.L., Barry, M.J. and Stoeckle, J.D. (1991), 'Advance Directives for Medical Care; a Case for Greater Use', *New England Journal of Medicine*, vol. 324, pp. 889–95.

Jennett, B. (2002), *The Vegetative State: Medical Facts, Ethical and Legal Dilemmas*, Cambridge University Press, Cambridge.

Official Solicitor to the Supreme Court (1996), 'Practice Note: Vegetative State', [1996] 2 FLR 375.

President's Commission (1983), *Deciding to Forego Life-sustaining Treatment*, Government Printing Office, Washington DC.

Rabkin, M.T., Gillerman, G. and Rice, J.D. (1976), 'Do Not Resuscitate Orders', *New England Journal of Medicine*, vol. 294, pp. 634–5.

Royal College of Physicians Working Group (1996), 'The Permanent Vegetative State', *Journal of Royal College of Physicians of London*, vol. 30, pp. 119–21.

Smedira, N.G., Evans, B.H., Grais, L.S., Cohen, N.H., Lo, B., Cook, M. et al. (1990), 'Withholding and Withdrawal of Life Support from the Critically Ill', *New England Journal of Medicine*, vol. 322, pp. 309–15.

Williams, B.T. (1989), 'Life-Sustaining Technology: Making the Decisions', *British Medical Journal*, vol. 298, p. 978.

Chapter 7

Autonomous Agency and Consent in the Treatment of the Terminally Ill

Susan L. Lowe

In their recent book, *The Case for Physician Assisted Suicide*, Sheila McLean and Alison Britton ask:[1]

> What could be more autonomous than a competent request for assistance, and what more disingenuous than to ignore autonomy simply because of the *nature* of the individual's clinical condition?

The claim here is that, when a person's clinical condition renders him incapable of acting on his own behalf to achieve some autonomously determined goal, refusing him assistance is quite simply failing to show him the respect which anyone should have for another person's autonomy. The rhetorical flourish with which the claim is made should, however, put us on our guard. In this chapter, I shall be taking a closer look at the concept of autonomy and its relation to the concept of consent. It will emerge that the concept of autonomy is more complex than is often assumed and that it readily gives rise to confusion. More importantly, I shall argue that attempts to justify either voluntary euthanasia or physician-assisted suicide in the name of patient autonomy are fundamentally misconceived. This is not to deny that it may be possible to justify these procedures on other grounds, though it is not my own opinion that they are in fact ever justifiable. My aim, rather, is to demonstrate that advocates of voluntary euthanasia and physician-assisted suicide present their opponents in a false light when they profess to justify those procedures in the name of patient autonomy. Patient autonomy is a value to which almost everyone would nowadays subscribe, so that to claim that it justifies voluntary euthanasia or physician-assisted suicide is to suggest that only a bigot could be opposed to those procedures. In this way, opponents of the procedures are cast in an unfavourable light by what is, in reality, nothing more than a show of rhetoric, which can at best be excused as arising from a misconception.

'Autonomy' is not, of course, a word that is often used in everyday conversation. It is a technical term used mainly by philosophers. There is a lot to be said for avoiding technical terms, as far as possible, when one is engaged in ethical analysis and argument, because they tend to obscure our vision. But the technical jargon of any discipline, including that of medicine, can be criticised on the same grounds. As has been well said by Anne Maclean, it is ironic to find writers on medical ethics avoiding medical jargon in the name of clarity, while at the same time indulging in bioethical jargon of their own.[2] Too often we see them using words such as 'beneficence' and 'maleficence', instead of the plain and simple phrases 'doing good' and 'doing harm'. And Maclean gives as another example the use of the word 'autonomy' when, she suggests, the phrase 'making up one's own mind' would do just as well. But *would* it do just as well, in fact? Sometimes, a technical term is needed because a complex concept has to be expressed, which is not easily captured in a simple phrase of everyday language. This appears to be the case with the word 'autonomy'.

We begin to see these complications when we examine the ways in which various philosophers have attempted to define the notion of autonomy. Here are two such attempts, drawn from recent books. First, in *The Blackwell Companion to Philosophy*, we find the following definition of an 'autonomous being':[3]

> An autonomous being is one that has the power of self-direction, possessing the ability to act as it decides, independent of the will of others and of other internal or external factors.

Then again, in his book *Philosophical Medical Ethics*, we find Raanan Gillon defining 'autonomy' thus:[4]

> Autonomy (literally, self-rule) is, in summary, the capacity to think, decide and act on the basis of such thought and decision freely and independently and without ... hindrance.

Gillon follows others in distinguishing between autonomy of thought, of will, and of action.[5] But, certainly, both he and the other source just quoted agree that the concept of *agency* – that is, the concept of a person's *being able to act* – is centrally involved in the concept of autonomy. In that case, autonomy is not, as Anne Maclean suggests, simply a matter of being able to 'make up one's own mind' about something, for that only involves, at most, autonomy of thought and of will or decision, not necessarily autonomy of action.

With these complications in mind, let us now proceed to cases. We can begin with some trivial, everyday examples, to help us to get our bearings without being distracted by weighty moral issues of life and death. One thing that we need to get clear about is that questions of autonomy are very sensitive to the distinction between acting and refraining from action. Another is that they become especially complex when more than one agent is involved.

Nothing, it may be thought, could more clearly illustrate the notion of a fully autonomous agent in action than a case such as the following. Feeling thirsty, I decide to stop work and make myself a cup of coffee. I go to the kitchen, make the coffee and drink it, thereby freely fulfilling my own autonomous decision. But things are immediately less simple when other agents are on the scene and undesired actions are in question. Suppose, for example, that I decide that I would like some coffee but don't want to stop work, so I ask you instead to make some coffee for me. Here my *decision* is still autonomous, but if I get the coffee it is not as a consequence of my own autonomous *action*, but rather of yours. And it would be preposterous for me to claim that you have an obligation to make me the coffee out of respect for my autonomy. You do nothing to restrict my *autonomy* by refusing to make me the coffee and, equally, nothing to enhance or effectuate it by complying with my request. My request depends for its fulfilment upon your *consent*, which, as an autonomous agent yourself, you are perfectly entitled to withhold. Indeed, a due respect for *your* autonomy would require me to acquiesce in your refusal. By the same token, however, if you offer me coffee which I don't wish to accept, I am perfectly entitled to turn it down. In this case, respect for *my* autonomy consists in your not making me accept something which, as a consequence of my own autonomous decision, I wish to refuse.

So here already we see an important difference, in cases involving more than one person, between requests for help and offers of assistance. Refusing a request for help – however justified or well-motivated that request may be – is perfectly consistent with showing due respect for the other person's autonomy, but imposing an offer of unwanted assistance is most certainly not. In both cases, however, what is crucial is the presence or absence of *consent* on the part of one of the people concerned. Imposing an offer of unwanted assistance on someone violates that person's autonomy, because it does not have his consent. Refusing a request for help does not violate the autonomy of the person making the request, because unless the person of whom the request is made consents freely to comply with it, his own autonomy is violated.

Now, it may be doubted whether these examples have any direct bearing upon cases involving the treatment of the terminally ill and more particularly upon the issue of voluntary euthanasia and physician-assisted suicide, because in such cases patients are rendered incapable, by their clinical condition, of performing certain actions which normal, healthy people are capable of performing for themselves. If someone refuses to comply with my request to make me a cup of coffee, I *can* still go and make it for myself, even though this will mean that I shall have to stop working, which I don't wish to do. Then I simply have to decide which I want more, to have some coffee or to go on working. I can decide autonomously and put my decision into effect autonomously. But, it may be said, mentally competent terminally ill patients already have their autonomy compromised. They are still able to make autonomous decisions, but they are prevented by their clinical condition from putting some of those decisions into effect, even though normal, healthy people would be able to carry out similar decisions autonomously.

The question, however, is whether this fact implies that to refuse such a patient's request for help in carrying out such a decision would, after all, constitute a failure to show due respect for the patient's autonomy. More particularly, does the fact that such a patient is incapable of carrying out for himself an autonomous decision to end his own life constitute a *relevant difference* between his requesting a physician to end his life, or to help him to end it, and my requesting you to make me a cup of coffee? If it does not constitute a relevant difference, then, just as your refusing my request could not be construed as restricting my autonomy and your complying with it could not be construed as enhancing or effectuating my autonomy, so, too, the physician's refusing or complying with the patient's request cannot be construed as having any such impact on the patient's autonomy – and consequently cannot be condemned or justified in the name of respect for the patient's autonomy. I shall argue that, indeed, there is *no* such relevant difference between the two cases and consequently that procedures such as voluntary euthanasia and physician-assisted suicide *cannot* be justified in the name of respect for patient autonomy.

At this point we should perhaps make clear, if it is not clear already, the precise distinction between voluntary euthanasia and physician-assisted suicide. Voluntary euthanasia is exemplified by the administration, by a physician, of a lethal injection, at the free or autonomous request of the patient. By contrast, in a case of physician-assisted suicide, the physician merely supplies the means whereby the patient may administer some lethal treatment to himself, taking account of the patient's particular physical disabilities. In an extreme case, this might be a matter of enabling the patient, by the mere flick of a switch, to initiate some process which he knows will result in his death. In both kinds of procedure, voluntary euthanasia and physician-assisted suicide, the physician plays an active part in the accomplishment of the patient's death. In both kinds of case, the physician performs actions which, if they were to be done in opposition to the patient's desires, would certainly be regarded as both immoral and illegal. And in both kinds of case, the physician performs actions which, if they were to be done at the freely made request of a perfectly healthy person, would also be regarded as both immoral and illegal, constituting, respectively, murder on the one hand and aiding and abetting suicide on the other. But the question which we have to address is whether, in the case of a terminally ill and physically incapacitated patient who freely requests a physician to perform such actions, the physician is actually under some moral obligation to comply with such a request, in the name of respect for the patient's autonomy. That the answer to this question is 'Yes' has certainly been suggested by some writers on medical ethics. Howard Brody, for example, says this:[6]

> Physicians ... have a moral obligation to respect the autonomous choices of their patients. Some few patients, even when provided with excellent palliative care, will autonomously select PAS [physician-assisted suicide] as their preferred option. Physicians should honour that request in these cases.

One consideration which might be thought to be relevant here is that, where mentally competent but terminally ill patients are concerned, it is generally accepted that their right *to refuse life-sustaining treatment* can be defended in the name of patient autonomy. Hence, it may be assumed that, by the same token, such a patient's right to be given a lethal injection or assistance in committing suicide can similarly be defended in the name of patient autonomy. However, as I have argued elsewhere,[7] no legitimate inference can be made from a right to refuse life-sustaining treatment to a right to be killed – and it can scarcely be denied that to be given a lethal injection or to commit suicide, with or without assistance, is to be killed. Hence, if any case is to be made for saying that voluntary euthanasia or physician-assisted suicide is justifiable in the name of patient autonomy, it must be made independently of any appeal to a patient's right to refuse life-sustaining treatment, even if the latter right can be defended in the name of patient autonomy. So let us see whether such an independent case can indeed be made. In my view, the answer to this question is decidedly 'No', for reasons which I hope to clarify further.

It cannot be denied, of course, that people who are physically incapacitated by terminal illness are severely disadvantaged, not least in respect of the range of autonomous actions which they can perform. They should consequently at least be accorded the rights which are accorded to disabled people in general. It is broadly accepted that society at large has a duty to see that disabled people should, as far as is possible, not be put at a disadvantage by their disability in comparison with able-bodied people. Hence, for example, it is right to require public buildings to be provided with wheelchair access and right to require employers not to discriminate unfairly against disabled applicants for jobs. However, we should acknowledge that society cannot necessarily *confer* autonomy, in respect of some range of activity, upon someone who has lost it. Sometimes, the most that can be achieved is to *compensate* someone for the loss of his autonomy of action, by conferring benefits upon him which he might normally be able to attain for himself. Providing wheelchair access to a public building will enable a disabled person to enter the building without anyone else's assistance, and to that extent does widen the range of autonomous action available to such a person. However, it is obviously impossible for a person to be *disabled* while also being *enabled* to perform the full range of autonomous actions open to a normal, able-bodied person. To restore such a full range of activity to a disabled person would be to cure him of his disability, which may not be possible. Thus, there are some kinds of autonomous action which a wheelchair-bound person simply *cannot* perform, such as running and jumping. The most that society can hope to do here is to compensate such a person for the loss of benefits imposed on him by this restriction of his range of autonomous action. Similarly, if a person's disability prevents him from doing paid work of any kind, the most that society can attempt to ensure is that he is compensated, as far as possible, for his loss of the benefits which paid work normally provides.

Now let us see how all of this bears upon the case of the terminally ill. One can readily envisage the following line of argument being advanced. The terminally ill patient is prevented, by his clinical condition, from performing certain autonomous actions which are available to normal, healthy people – notably, the action of suicide. Therefore, society should either take steps to enable these people to commit suicide, if it is their autonomous decision to do so, or else – if that is impossible – it should compensate these people for their inability to commit suicide by conferring upon them the benefits which, as they see it, suicide would bring them. In the first case, then, society should make physician-assisted suicide available, where possible, to the terminally ill. And, secondly, in the case of terminally ill patients for whom this option is impossible, society should make voluntary euthanasia available. Moreover, it may be contended, these obligations are incurred by society *simply on the grounds of respect for individual autonomy*.[8] The obligations in question, it may be held, are entirely comparable with those which society accepts, on the same grounds, in respect of disabled people in general.

However, there is a serious flaw in this argument. Society only incurs an obligation to remove obstacles to the autonomous agency of disabled people, or to compensate them for the loss of benefits which their disabilities cause, when it is recognised that people quite generally have a *right* to act in the ways in question and an *entitlement* to the corresponding benefits. Thus, it is widely accepted that people have a right to work and an entitlement to the benefits which paid employment normally brings. That is why, when someone is prevented from exercising that right in virtue of having a diminished range of autonomous action available to him, society recognises an obligation to make amends, either by enabling him to exercise that right by removing certain obstacles in his way or else by compensating him for the consequent loss of benefits. But not every action which is legally permissible is one which people have a *right* to engage in. For example, it is legally permissible to get drunk, at least in the privacy of one's own home, but no one has a *right* to do so. Similarly, adultery is, in Britain at least, legally permissible, in that it is not a criminal or civil offence. But that doesn't imply that people have a *right* to commit adultery, which other members of society ought to respect. There is no good reason why society should recognise that its members have a right to perform actions of a certain kind – even if it deems that such actions should be legally permissible – unless it can be persuaded that benefits accrue from such actions to which all of its members are entitled. It may deem that certain intrinsically harmful actions should be legally permissible because to criminalise them would be likely to create still more harm. In the case of such actions, it is clear that society does not and need not recognise that its members have a right to engage in them.

Now, as in the cases of drunkenness and adultery, it is legally permissible in Britain to commit suicide, in that it is neither a criminal nor a civil offence. But, for the reason just given, this doesn't imply that people have a *right* to commit suicide, which other members of society ought to respect. As Yale Kamisar has justly pointed out,[9]

> The fact that we no longer punish suicide or attempted suicide does not mean that we *approve* of these acts or that we *recognize* that an individual's right to 'self-determination'or 'personal autonomy' extends this far.

Just as in the cases of drunkenness and adultery, so too in the case of suicide, many members of society hold that this is an action which, while legally permissible and rightly so, is deeply morally repugnant. And such people are fully entitled to that opinion, which deserves to be respected by other members of society, even if they do not share it themselves.

Of course, the terminally ill patient may believe that he is deprived of what he sees as a benefit – death – in virtue of his incapacity to commit suicide unaided. And if it could be established, to the satisfaction of society at large – including the physicians concerned – that death in such circumstances *would* be a benefit to the patient, then indeed it might be argued that the patient has a right to it and that society has an obligation to make that benefit available to him, either through physician-assisted suicide or through voluntary euthanasia. However, this would be to argue for the legitimacy of physician-assisted suicide and voluntary euthanasia on grounds quite other than that of the requirement to respect individual autonomy. Society has no obligation to regard as a benefit whatever any particular member of society may happen to regard as a benefit, even if that member of society has arrived at his opinion through a process of autonomous thought and decision. We may respect that person's opinion and agree that he has a perfect right to hold it. But that doesn't mean that we should feel obliged to help that person to attain the supposed benefit if circumstances should arise which prevent him from attaining it for himself, by his own autonomous action. Respect for him as an autonomous moral being does not carry with it any such obligation.

An important factor which we should reintroduce here is that of *consent*. A request for help depends for its satisfaction upon the consent of the person who is being asked to help. Sometimes, indeed, we are under an obligation to give our consent to such a request. But in other cases, and much more frequently, to insist upon our consent would constitute a violation of our own autonomy. And this, very arguably, is how matters stand in the cases of physician-assisted suicide and voluntary euthanasia. Far from it being true that these procedures can be justified in the name of respect for patient autonomy, it can be argued that for society or patients to impose any pressure on physicians to engage in such activities would itself constitute a failure to show proper respect for the *physicians'* autonomy and, more particularly, for their right to act autonomously in accordance with the dictates of their own conscience. Even if physician-assisted suicide and voluntary euthanasia were legalised, the consent of physicians to engage in it would have to be freely given as a consequence of their own autonomous deliberation and decision.

To conclude, then, neither physician-assisted suicide nor voluntary euthanasia can be justified purely on the grounds of the need to respect patient autonomy. The incapacity of the patient to bring about his own death unaided only becomes a relevant factor if it can already be argued, *on other grounds*, that this restriction on the scope of his autonomous action deprives him of rights and benefits to which he is entitled as an equal member of society. Moreover, to the extent that the consent of the physician would be required for both procedures, the obligation to respect individual autonomy in fact implies that physicians should not be put under any pressure to carry out such procedures, even if they were to be legalised.

Notes

1 McLean and Britton, 1997, p. 20.
2 Maclean, 1993, p. 200.
3 Bunnin and Tsui-James, 1996, p. 743.
4 Gillon, 1992, p. 60.
5 Ibid., p. 61.
6 Brody, 1997, p. 137.
7 Lowe, 1997, pp. 154–8.
8 Just such a view seems to be expressed by Harris, 1985, pp. 204–5.
9 Kamisar, 1995, p. 229.

References

Brody, H. (1997), 'Assisting in Patient Suicides *is* an Acceptable Practice for Physicians', in Weir, R.F. (ed.), *Physician-Assisted Suicide*, Indiana University Press, Bloomington and Indianapolis.

Bunnin, N. and Tsui-James, E.P. (eds) (1996), *The Blackwell Companion to Philosophy*, Blackwell, Oxford.

Gillon, R. (1992), *Philosophical Medical Ethics*, Wiley, London.

Harris, J. (1985), *The Value of Life*, Routledge and Kegan Paul, London.

Kamisar, Y. (1995), 'Physician-Assisted Suicide: The Last Bridge to Active Voluntary Euthanasia', in Keown, J. (ed.), *Euthanasia Examined*, Cambridge University Press, Cambridge.

Lowe, S.L. (1997), 'The Right to Refuse Treatment is not a Right to be Killed', *Journal of Medical Ethics*, vol. 23, pp. 154–8.

Maclean, A. (1993), *The Elimination of Morality*, Routledge, London.

McLean, S. and Britton, A. (1997), *The Case for Physician Assisted Suicide*, Harper Collins, London.

Chapter 8

The 'Frankensteinian' Nature of Biotechnology

David E. Cooper

Introduction

The revolution in biotechnology, especially – but not only – in its medical applications, doubtlessly generates a host of particular questions of moral and legal interest. One thinks, for example, of such issues as the ownership of genes and the morally relevant distinction, if any, between somatic cell and germ-line therapies. My concern in this chapter, however, is with something more general: the whole moral psychology, one might say, which is indicated by the anxious and hostile responses which many people display towards genetic engineering and other interventionist biological procedures.

Much effort is currently devoted to educating a suspicious public about the nature of and prospects for biotechnology, in the confidence that, thus educated, suspicion and fears will be allayed. Some of this effort, no doubt, is an honest contribution to rational resolution of what Hans Blumenberg called the tension between 'uneasiness about science's autonomous industry and the constraints resulting from its indispensability' (Blumenberg, 1966: 230). Some of it, though, gives the impression of a disingenuous attempt to hold the fort until the public simply gets accustomed to biotechnology as a mundane feature of tomorrow's world – an attempt which might invite Dostoievsky's remark, in *Crime and Punishment*, that 'Man gets used to everything – the beast!' Whether or not people will be educated out of, or simply grow out of, their suspicion and fears depends on what these are. If, as spokesmen for biotechnology often insist, they are the products of 'irrational taboos', pig ignorance, Luddite nostalgia, or the 'genetic pornography' put about by the popular media (quoted in Turney, 1998: 3), then educative effort or the sands of time should do the trick. But if, as I judge, such diagnoses are superficial, the prospects for 'winning over' the public are much less certain.

Frankenstein and its Lessons

To indicate the superficiality of these diagnoses and to prepare for a less superficial one, I shall invoke a name that has invariably been invoked in debates about interventionist biology during the century just passed – that of Frankenstein. The titles of two recent books on the ethical and social aspects of biotechnology, *The Frankenstein Syndrome* (Rollin, 1995) and *Frankenstein's Footsteps* (Turney, 1998), nicely confirm how, in the words of one of the authors, the story of Frankenstein has been 'the governing myth of modern biology' (Turney, 1998: 3). The name of the famous creator of a monster – sometimes transferred to the monster itself – is ubiquitous in modern debate. James B. Watson, for example, complained about the 'left-wing nuts and environmental kooks [who] have been screaming that we will create some kind of Frankenstein bug' (quoted in Kolata, 1997: 7). During the great debate twenty years ago over recombinant DNA research at Harvard, the *Washington Star* asked 'Is Harvard the proper place for Frankenstein tinkering?', while the Mayor of Cambridge, Massachussetts, promised to prevent 'Frankenstein monsters crawling out of the sewers'. One Harvard historian of science gave the title 'Frankenstein at Harvard' to his essay on the affair (quoted in Turney, 1998: 196–7).

As the remark by Watson helps to testify, an interesting change seems to have occurred over recent years in the purpose for which the name gets invoked. Time was when it was most likely to be heard from the lips of *critics* of technical innovations in medicine, the offending innovators being accused of emulating Victor Frankenstein. ('Victor', for some reason, became 'Henry' in the film versions.) Nowadays, it is more commonly heard from the lips of *advocates* of biotechnology when accusing their critics of being in the grip of myth and science fiction. One contributor to the British parliamentary debate on the Human Fertilization and Embryology Bill, for example, complained of the way that the 'spectre of Dr Frankenstein's monster impinges heavily on our subconscious when we address ourselves to the problem of embryology, causing a fear and revulsion' (quoted in Turney, 1998: 248). Perhaps it does so impinge, but it belongs to what I was calling the 'superficial diagnosis' to conclude that such 'fear and revulsion' should, in consequence, be lightly dismissed.

It is certainly not my intention to add to the long list of interpretations of Mary Shelley's novel – to, say, the Marxist one which interprets the Doctor as the capitalist class and the monster as the proletariat which rises up against the class which produced it, or the feminist one which reads the tale as a diatribe against the '"incestuous" violation of life' by 'penetrative' male science (Hindle, 1992: xxxvii ff). But I do want to bring out a number of features of the story, ones that are often overlooked, which enable us to identify the character of the 'fear and revulsion' which many people experience in the face of interventionist biology. These features teach, so to speak, a number of interlocking lessons applicable, I suggest, in the current state of debate.

First, we should bring out a feature, not so much of the novel itself, but of its place in the imagination over the last two centuries. Frankenstein does not simply intervene in nature, but specifically in the generative process of (human) life. He creates a creature by artificial means. Now it is surely no accident, nor a matter of its literary merit, that this story – rather than ones about 'mad scientists' who intervene in the world's weather or the ocean's currents – has become *the* paradigmatic myth of questionable intervention in nature. In my judgement, it is mistaken to regard the myth as signifying a quite general fear of scientific innovation or a global Luddite reluctance towards technological advance. Only the development of nuclear power has rivalled biological interventionism in its capacity to elicit 'fear and revulsion': but, as the many fictions devoted to the nuclear threat themselves demonstrate, the main focus of such fear has not been on annihilation or devastation, but on something itself biological – the effect of radiation on the body. An unseen invasion of the body, it seems, makes for more compulsive reading or viewing than its incineration.

Confirmation of this point is provided by the outcry, especially in the United Kingdom, against genetically modified foods. This does not testify to a general hostility towards technological intervention in agriculture: few people, after all, object to the use of sophisticated machinery for growing and marketing food. It testifies, rather, to a sense – rarely given cogent expression – that the production of GM foods represents a questionable, quite direct intervention in the life process of animals and plants, and hence only a slightly less direct intervention in the life processes of ourselves, the consumers of the modified organisms. The first lesson, then, is that those spokesmen for biotechnology who pass off resistance to it as a function of a general hostility to science miss the point. There really is something special about the misgivings people have about biological techniques like cloning, germ-line therapy, and genetic modification, and these would remain even if science at large enjoyed a rosier reputation than it does.

So, to help identify the character of those misgivings, let me mention some instructive features of *Frankenstein*, briefly indicating the lessons suggested, and then going on to elaborate them. Here, to introduce the first feature, is how Victor Frankenstein describes his response, on that 'dreary night of November', when he 'beheld the accomplishment of [his] toils':

> I had worked hard for nearly two years, for the sole purpose of infusing life into an inanimate body … I had desired it with an ardour that far exceeded moderation; but now that I had finished … breathless horror and disgust filled my heart. Unable to endure the aspect of the being I had created, I rushed out of the room. (Shelley, 1992: 56)

People who know only the cinema versions often assume that the point of the story is to warn against the terrible *consequences* that intervention in natural processes may have. But Victor's 'horror and disgust' are *immediate*, and so prior to the murderous deeds his creature – not without some justification – later commits. Mary Shelley makes it plain in her Preface, moreover, that what is 'supremely frightful' is simply the 'human endeavour to mock the stupendous mechanism of the Creator of the world', *irrespective* of the later consequences of that endeavour (Shelley, 1992: 9). So here is a second lesson: perhaps the 'fear and revulsion' that many people experience towards genetic engineering and its relatives are not, as often supposed, primarily pragmatic ones concerning disastrous results.

Two further, related features of the novel are these. First, its central character, unlike that of several movie versions, is not the creature, but Dr Frankenstein himself. Second, a discernible theme is that of science as a force or enterprise that takes over and drives the individual scientist. Both Frankenstein and the scarcely less obsessed narrator, Walton, attest to the 'intoxicating draught' or 'genius' of natural philosophy that has 'regulated [his] fate' (Shelley, 1992: 27). The lesson of these two features, I suggest, is that moral concern is aroused not only – perhaps not mainly – about what scientists do and the consequences of what they do, but about the character both of science as an 'autonomous industry' and of the scientists themselves, taken over and 'regulated' by its onward drive.

A final feature concerns the intellectual background against which the novel, first published in 1818, was written. Through her father, William Godwin, and her husband-to-be, the poet Percy Bysshe Shelley, Mary Shelley (1797–1851) was well-versed in the scientific and philosophical debates of her age. She was, for example, acquainted with the work and theoretical outlooks of Erasmus Darwin (1731–1801) and Humphry Davy (1778–1829), whose words Frankenstein sometimes apes. More to the point, she knew the arguments raging between Enlightenment rationalists and their Romantic critics, arguments centring on the viability of a purely mechanistic account of the world and human beings. By assembling a living creature from bodily bits and pieces, Victor clinches the case, seemingly, for the mechanists and against those Romantics for whom a human being, a human body, is the epitome of an 'organism' that defies analysis and explanation in mechanistic terms. It is not then far-fetched to regard, as a further lesson of Victor's horror and revulsion, the profound injury, the 'metaphysical horror', caused by the extension of the mechanistic world-view to the domain of the human or, more widely, the living.

The 'Yuk' Factor

I now want to elaborate on these lessons in relation to current moral debates about biotechnology, beginning with that drawn from the immediacy of Victor's horror and revulsion. 'The strength of the "yuk" factor', one author rightly states, 'when the traditional order of things comes under threat' – irrespective of the consequences of that threat – 'has been greatly underestimated' (Johnson, 1998: 141). Doubtless there is a strategic advantage for enthusiasts of biotechnology in trying to centre debate on the potential outcomes of their work: for they are then on home ground, able to pit their expert predictions against the wild imaginings – of cloned Hitlers, blackmarkets in spare bodily parts, unstoppable GM 'superweeds', and the like – indulged in by ignorant laymen and 'genetic pornographers'. But, if I am right, construing the debate in these consequentialist, purely pragmatic terms is illegitimately to ignore, or trivialise, 'the "yuk" factor'.

Some writers, it is true, take account of – and do not simply dismiss – what they are nevertheless apt to describe as 'irrational taboos', 'ancient theological scruples', or 'mere prejudices' against interventionist biology. But they do so, typically, from a strangely disengaged perspective, treating them as expressive of feelings which need not be taken seriously except for purposes of utilitarian calculation. As one writer puts it, gene-splicing and the like are 'morally objectionable only if people's feelings are strong enough to outweigh the expected net utility of employing the technique' (Häyry, 1994: 208).[1] Even Mary Warnock, who attributes a more important place to 'feelings' in shaping an acceptable morality, writes as if their importance is only the pragmatic one of standing in the way of workable legislation. 'It is of no use to permit something to be done without control if th[e] collective voice [of 'private moral sentiments'] too deeply abhors it' (Warnock, 1998: 22). The validity or otherwise of the 'sentiments' is not, it seems, worth examining, only the obstacle they may present to policy-makers.

Such attitudes amount less to underestimation of 'the "yuk" factor' than to refusal even to reflect on what revulsion at, say, cloning or at engineering animals with organs transplantable to humans might signify. As such, a crucial element in criticism of such procedures is simply by-passed. But what, then, do expressions of 'yuk' signify if not mere 'irrational taboos' and 'ancient theological scruples'?[2] Here we need to turn to the other lessons drawn from my brief discussion of the Frankenstein story.

One lesson, I suggested, was that moral concern may be directed more towards the agents of scientific innovations than at the innovations themselves. Crudely put, what seems wrong about such innovations is that they are the doings of a certain kind of person, caught up in a certain kind of enterprise. It is striking how many of the most effective works of science fiction are dominated by a central, charismatic 'mad scientist' – H.G. Wells' Dr Moreau, for example, and Victor Frankenstein himself. It is striking, too, that the disturbing gospels pronounced by such protagonists are clearly intended to echo those of real-life scientists. Triumphant references to science 'bestowing upon man powers to ... modify the beings surrounding him', to 'interrogate nature' and be its 'master', might have been Victor's, but were actually Sir Humphry Davy's (quoted in Hindle, 1992: xxvi). Moreau's remorseless detachment from the sufferings of his laboratory animals is that of the French physiologist Claude Bernard (1813-78), who speaks of the scientist as one 'possessed. He doesn't hear the cries of the animals, he does not see their flowing blood ... [He] is aware of nothing but organisms which conceal from him the problems he is wishing to solve' (quoted in Turney, 1998: 158). And doubtless there are characters in contemporary science fiction who echo one Nobel prize-winning biologist's, Paul Berg's, pronouncement that he would stop his work 'if there were a sound practical reason, but not if it were an ethical judgement' (quoted in Bains, 1990: 3).

It should not, surely, cause surprise if the objects of the 'yuk' response are often the men, the scientists, who can come out with such pronouncements. There is, many people will feel, something wrong with such men, over and above their deeds and the consequences of these. And if it sounds odd, in our present moral climate, to judge people in such an unpragmatic way, we should recall that, for centuries, the predominant climate of moral philosophy was that of so-called 'virtue ethics', an approach whose ancestry goes back to Aristotle. The crucial questions to ask, in such an ethics, were ones like 'What kind of person is he?' and 'Does he possess the virtues?', with judgement on the actions the person performs being shaped by the answers to those questions. Very roughly, a bad action was one a bad person would perform, not vice-versa.

Or, to sound less 'moralistic' and more 'modern', the idea was and, for some people, still is, that we should guide our judgement on what people do and propose, scientists included, by reflecting on the acceptability or otherwise of the deep perspective – the 'mind-set' or *Weltanschauung* – which their deeds and proposals may manifest.

Life and Reductionism

What, then, might be the 'vice' or the distorted *Weltanschauung* which people claim to discern among interventionist scientists like those quoted and which gives rise to 'the "yuk" factor'? There is much truth, I suspect, in an answer implicit in Iris Murdoch's notion of 'humility', understood not as some Uriah Heep-ish self-abasement, but as a 'selfless respect for reality', for 'the independent existence' of the creatures and things that make up reality (Murdoch, 1970: 85, 95).[3] What is wrong with those interventionist scientists is what was wrong with Prometheus, Faust, Frankenstein, Moreau and a host of other legendary and doomed figures – hubris. I shall return to that thought, but only after exploring a slightly different answer – one which, moreover, honours my earlier point that misgivings about biotechnology are 'special' and not a mere function of a general suspicion of the hubris of 'masterful' science.

My answer also exploits another lesson drawn from the discussion of *Frankenstein* and the philosophical debates lurking in the background to that book – the misgivings or revulsion which can be aroused by extending a mechanistic *Weltanschauung* to the human domain. Richard Dawkins tells us that organisms, including human bodies, are 'merely vehicles for genes' (quoted in Holland, 1998: 229). Writing of the so-called 'genetic body' – the one 'sought by the human genome project' – another writer reports the increasingly common view that 'we are what our genes tell us we are ... [the genetic body] is who we are ... the centre of [our] identity' (quoted in Turney, 1998: 218-9). Claims like these no doubt go beyond what is strictly required of a biological theory in order for it to underpin genetic engineering. But they are claims to which many biologists seem prone these days, and they are ones which easily encourage *insouciance* towards both species and individual organisms. They amount to a genetic reductionism which, in one biologist's words, 'makes it legitimate to shunt genes around from one species to any other species' (quoted in Holland, 1998: 236), and prompts another to wonder why people make such a fuss about transgenic procedures given that 'all genetic material is the same, from worms to humans' (quoted in Linzey, 1994: 150–51).

It is unnecessary, I think, to dwell on the fact that utterances like these display no more conceptual acumen or sensitivity than that of someone who, having observed that all musical raw material is the same, from rugby songs to the B-minor Mass, concludes that it is perfectly legitimate to shunt notes around from one composition to any other. Anyway, it is not in my brief to spell out the confusions of genetic reductionism. What is in my brief is to stress that it is surely such reductionist claims, perceived as insensitive, indicative of an obsessively narrow vision of life, and responsible for a culpable *insouciance* towards species and individuals – now treated as mere vehicles for genes and their transfer – which incites hostility, even revulsion, towards the purveyors of the *Weltanschauung* to which such claims lend expression.

One aspect of that *insouciance* is worth dwelling on. On his first visit to Europe, what most shocked Gandhi, apparently, was the casual willingness of Westerners to hand over their bodies to the control and offices of others, doctors in particular. But perhaps Gandhi underestimated the degree to which even Westerners, at least since the atrophy of belief in the soul, identify themselves with their bodies, and hence will permit only a limited interference with them for circumscribed purposes. We do not so much *have* bodies as *are* those bodies. The bodies in question, however, are, to borrow from the vocabulary of phenomenology, 'lived bodies' – the vehicles or expressions of our lives, relations with one another, convictions and feelings, and comportment towards the world.[4] They are not the bodies of anatomy and biology – vehicles or containers for little bits of them, corpuscles, genes, or whatever. Genetic reductionism, with its cheerful corollary of 'shunting genes around', amounts to an elision of the distinction between the 'lived' and the biological body, and hence to an erasure of our notion of the individual person. Those who understandably recoil from that erasure are unlikely to be soothed by the remark of one advocate of cloning that 'most people don't have any individuality anyway' (quoted in Kolata, 1997: 33).

Two predictable replies will be made to what I have been saying. First, it will be urged that, in the light of recent advances in biology, surely we *ought* to erase traditional notions of the individual person and the body he or she is – to replace them, perhaps, by that of 'selfish genes' and those unsuspecting hosts whom, it will one day be clear, it was merely quaint to think of as 'persons'. The answer to that proposal is that these traditional notions are *ineliminable*. Only outside of the flow of everyday life, of the *Lebenswelt*, is it possible to suspend the concepts which play an indispensable role in our understanding and explanations of, in our attitudes and natural responses to, human beings – and indeed many animals, too.

Second, it will be said that genetic techniques are but an extension of the interventions into our bodies to which, as Gandhi noted, we Westerners have for long been willingly accustomed. But as I noted, when citing Gandhi, we accept these only for circumscribed purposes. As John Turney puts it, we accept interventions 'in order to maintain the integrity of the body', not in order to produce 'deliberately engineered beings' (Turney, 1998: 219). This is why, I suspect, we cheerfully swallow Viagra and Prozac, but baulk at cloning and even the ingestion of GM foods. Doubtless the boundary between maintaining integrity and engineering is not sharply drawn, but to suppose that there are not two distinguishable territories here, however contested the border, would display just that insensitivity to ordinary moral psychology which, I have been arguing, is all too familiar in attempts to disabuse people of their suspicion of interventionist biology.

Integrity

The term 'integrity', as just used by Turney, crops up regularly in discussions of the matters with which I have been dealing. It points to something akin to Iris Murdoch's 'humility'. Arguably, the disquiet expressed by invocations of the name of Frankenstein might be best diagnosed as a sense that the integrity of things – living things, at least – is violated, indeed doubly violated, by the attitudes which often lurk behind the enterprise of biotechnology. The integrity of a living thing – what makes it the distinctive being it is – is diluted and distorted when it is regarded only in its relationship to, its potential use in, human endeavours and ambitions. And it is all but eliminated when the thing is pared down to the stuff – the genetic material – of which it is composed, with the result that the only distinctiveness it has is the superficial one of being a certain vehicle for this stuff. To regard the natural world, therefore, as so much stuff conveniently divided up into living things which are at our disposal is doubly to violate the integrity of things.

That there is this widespread sense of violated integrity when the living things happen to be human bodies is obvious. A human body is not simply a vehicle of its genes, but the vehicle of a person's engagement with the world and other people. To regard the body in the reductionist manner is to diminish persons and their distinctiveness. This is something which champions of 'hi-tech', interventionist medicine ignore at their peril. It is one reason, for example, why the prospect of engineering human or human-like bodies to serve as 'spare parts' emporia for transplant operations is one with the power to appall. The sense of violated integrity is only slightly less pronounced when the living things are non-human animals. Advocates of transgenic procedures like to say that these procedures are only the speedy and efficient version of breeding techniques which go back to the dawn of husbandry. But Stephen Clark is surely right to observe a difference between 'old-style pastoralists', who had to honour the natural ends of animals in order to profit from them, and 'new-style artificers', who first 'work out where the profit lies, and mould the ends to suit them' (Clark, 1997: 71). For example, it is one thing, surely, to exploit the natural virility and aggression of a bull in rearing a herd, another thing to genetically engineer a peculiarly placid bull with 'the bovine equivalent of Down's Syndrome, replacing the cantankerous animal [of] today with a placid ... fat animal' easy to control and extract semen from (Bains, 1990: 184).

Where the living things are neither sentient nor intelligent, the sense of violated integrity – at the hands of the GM food industry, say – may be less widespread: but it is surely there – and understandably so. I am reminded of a remark by Ludwig Wittgenstein. These days, he said, anyone is regarded as 'stupid' who doesn't know that water is simply H_20: and thereby, he adds, 'the most important questions are concealed' (Wittgenstein 1980: 71). Water, he is implying, is not *just* H_20: it is what sustains life, pounds on to the beach, what we swim in and baptise with. To reduce water to a neutral stuff at our disposal is to be blind to what it is, to its integrity. As it stands, Wittgenstein's complaint is more akin to the kind of despair which Martin Heidegger felt at the reduction, as he saw it, of the Rhine to a power supplier (Heidegger, 1977) than to qualms at GM agriculture. But, to make his point, Wittgenstein's example could as well have been of wheat, rice or – something he mentions in the same passage – sugar, as of water. They, too, are not *just* XYZ or whatever their chemical descriptions specify.

People respond differently to this talk of biotechnology's violation of the integrity of things, of its lack of humility towards life. In the atmosphere of debate which prevails on radio or television, it would doubtlessly be mauled by some terrier-like interviewer. Elsewhere it would receive a more sympathetic hearing. My argument is not that we must wholeheartedly endorse that sense of violation and so condemn biotechnology as the devil's work. It is, rather, that we should appreciate both what that sense is and that it is one which is shared, however inchoately, by many people. Unless these are appreciated – unless the 'yuk' factor is given its proper and serious due – discussion of such important issues as cloning, transgenic procedures and GM foods will inevitably continue in the same shrill, vituperative and unhelpful atmosphere which currently surrounds it.

Notes

1 Similar points are made by Harris, 1994, and Glover, 1984.
2 The expression 'ancient theological scruples' is from an open letter signed by Bernard Crick, Richard Dawkins and other members of The International Academy of Humanists in which they complain of a 'Luddite rejection of cloning' (quoted in Kolata, 1997: 97).
3 On the theme of humility and integrity, see also Cooper, 1998, and Cooper, 2001.
4 See, especially, Merleau-Ponty, 1962.

References

Bains, W. (1990), *Genetic Engineering for Almost Everybody*, Penguin, Harmondsworth.
Blumenberg, H. (1966), *The Legitimacy of the Modern Age*, MIT Press, Cambridge, Mass.
Clark, S.R.L. (1997), 'Natural Integrity and Biotechnology', in Laing, J. and Oderberg, D. (eds), *Human Lives*, Macmillan, London, pp. 58–76.
Cooper, D.E. (1998), 'Intervention, humility and animal integrity', in Holland, A. and Johnson, A. (eds), *Animal Biotechnology and Ethics*, Chapman & Hall, London, pp. 145–55.
Cooper, D.E. (2001), 'Philosophy, Environment and Technology', in O'Hear, A. (ed.), *Philosophy at the Millennium*, Royal Institute of Philosophy Lectures, Cambridge University Press, Cambridge.
Glover, J. (1984), *What Sort of People Should There Be?*, Penguin, Harmondsworth.
Harris, J. (1994), 'Biotechnology, Friend or Foe?', in Dyson, A. and Harris, J. (eds), *Ethics and Biotechnology*, Routledge, London.
Häyry, M. (1994), 'Categorical Objections to Genetic Engineering – a Critique', in Dyson, A. and Harris, J. (eds), *Ethics and Biotechnology*, Routledge, London, pp. 202–15.
Heidegger, M. (1977), *The Question Concerning Technology and Other Essays*, Harper & Row, New York.
Hindle, M. (1992), Introduction to Mary Shelley's *Frankenstein: Or The Modern Prometheus*, Penguin, Harmondsworth.
Holland, A. (1998), 'Species are Dead. Long Live Genes!', in Holland, A. and Johnson, A. (eds), *Animal Biotechnology and Ethics*, Chapman & Hall, London, pp. 225–42.

Johnson, A. (1998), 'Needs, Fears and Fantasies', in Holland, A. and Johnson, A. (eds), *Animal Biotechnology and Ethics*, Chapman & Hall, London, pp. 133–44.
Kolata, G. (1997), *Clone: The Road to Dolly and The Road Ahead*, Penguin, Harmondsworth.
Linzey, A. (1994), *Animal Theology*, SCM Press, London.
Merleau-Ponty, M. (1962), *Phenomenology of Perception*, Routledge & Kegan Paul, London.
Murdoch, I. (1970), *The Sovereignty of Good*, Routledge & Kegan Paul, London.
Rollin, B. (1995), *The Frankenstein Syndrome: Ethical and Social Issues in the Genetic Engineering of Animals*, Cambridge University Press, Cambridge.
Shelley, M. (1992), *Frankenstein: Or The Modern Prometheus*, Penguin, Harmondsworth.
Turney, J. (1998), *Frankenstein's Footsteps: Science, Genetics, and Popular Culture*, Yale University Press, New Haven.
Warnock, M. (1998), 'A Path to the Permissible', *Times Higher Education Supplement*, 29 May 1998, p. 22.
Wittgenstein, L. (1980), *Culture and Value*, Blackwell, Oxford.

Index

Printed and bound by CPI Group (UK) Ltd, Croydon, CR0 4YY
22/10/2024
01777620-0014